Springer Series in Reliability Engineering

Series Editor

Hoang Pham, Industrial and Systems Engineering, Rutgers, The State University of New Jersey, Piscataway, NJ, USA

Today's modern systems have become increasingly complex to design and build, while the demand for reliability and cost effective development continues. Reliability is one of the most important attributes in all these systems, including aerospace applications, real-time control, medical applications, defense systems, human decision-making, and home-security products. Growing international competition has increased the need for all designers, managers, practitioners, scientists and engineers to ensure a level of reliability of their product before release at the lowest cost. The interest in reliability has been growing in recent years and this trend will continue during the next decade and beyond.

The Springer Series in Reliability Engineering publishes books, monographs and edited volumes in important areas of current theoretical research development in reliability and in areas that attempt to bridge the gap between theory and application in areas of interest to practitioners in industry, laboratories, business, and government.

Now with 100 volumes!

Indexed in Scopus and EI Compendex

Interested authors should contact the series editor, Hoang Pham, Department of Industrial and Systems Engineering, Rutgers University, Piscataway, NJ 08854, USA. Email: **hopham@soe.rutgers.edu, or Anthony Doyle, Executive Editor, Springer, London. Email: anthony.doyle@springer.com**.

Kim Phuc Tran
Editor

Artificial Intelligence for Safety and Reliability Engineering

Methods, Applications, and Challenges

Springer

Editor
Kim Phuc Tran
ULR 2461 - GEMTEX - Génie et
Matériaux Textiles
ENSAIT and Graduate School
MADIS-631, University of Lille
Lille, France

ISSN 1614-7839 ISSN 2196-999X (electronic)
Springer Series in Reliability Engineering
ISBN 978-3-031-71494-8 ISBN 978-3-031-71495-5 (eBook)
https://doi.org/10.1007/978-3-031-71495-5

This Springer imprint is published by the registered company Springer Nature Switzerland AG
The registered company address is: Gewerbestrasse 11, 6330 Cham, Switzerland

Contents

Contents

Introduction to Artificial Intelligence for Safety and Reliability Engineering

Kim Phuc Tran

Abstract This chapter offers an overview of the book about Artificial Intelligence (AI) for Safety and Reliability Engineering within Industry 5.0, setting the stage for detailed discussions in subsequent chapters. AI can deliver novel and more accurate insights compared to traditional methods in the field of Safety and Reliability Engineering. Additionally, this book addresses the importance of ensuring the safety and reliability of AI systems, which are integral to the Industry 5.0 revolution. We will introduce critical topics that will be explored in the following chapters of the book.

1 Introduction

The Industrial Revolution 4.0 and 5.0 have profoundly transformed smart factories through the widespread adoption of technologies such as the Internet of Things (IoT), Human-centered Artificial Intelligence (HAI), Big Data, and interaction between humans and machines [1–5]. Several pieces of research in the literature regarding the application of Artificial Intelligence (AI) to ensure industrial systems' reliability, maintainability, and safety [6–9]. Moreover, ensuring the Safety and Reliability of AI systems is an important mission in building trustworthy AI systems [10]. However, there is currently, no reference book presenting systematic AI for Safety and Reliability Engineering and ensuring the Safety and Reliability of AI systems. This book addresses the safety and reliability engineering issues of real industrial systems with AI such as maintenance decision-making and risk, Safety and Reliability of AI systems, Predictive maintenance (PdM), optimization of reliability and safety aspects, etc.

K. P. Tran (✉)
ULR 2461 - GEMTEX - Génie et Matériaux Textiles, ENSAIT, University of Lille, 59000 Lille, France
e-mail: kim-phuc.tran@ensait.fr

International Chair in DS & XAI, International Research Institute for Artificial Intelligence and Data Science, Dong A University, Danang, Vietnam

K. P. Tran (ed.), *Artificial Intelligence for Safety and Reliability Engineering*, Springer Series in Reliability Engineering,
https://doi.org/10.1007/978-3-031-71495-5_1

This book introduces readers to the latest advancements in theoretical research, technological development, and applications of AI for Safety and Reliability Engineering, as well as the safety and reliability of AI systems, addressing current challenges and opportunities in Smart Manufacturing. It explores technologies to ensure the safety and reliability of AI in industrial applications and provides ready-to-use algorithms for various potential Smart Manufacturing scenarios. Each chapter includes case studies to help readers and practitioners apply these tools to real-world Smart Manufacturing processes. Additionally, the book offers insights into preserving data privacy and security in Smart Factories by designing a Federated Learning-based architecture.

2 Main Features of This Book

The key features of this book are given as follows:

1. A survey on Safety and Reliability of Artificial Intelligence systems
2. A review and perspective on Physics-informed Machine Learning for Industrial Reliability and Safety Engineering
3. Predictive Maintenance Optimization based on Genetic Algorithms for Future Industrial Systems
4. Anomaly Detection for Catalyzing Operational Excellence in Complex Manufacturing Processes
5. Human-Centered Edge Artificial Intelligence for Smart Factory Applications in Industry 5.0
6. Human-Centered Edge AI and Wearable Technology for Workplace Health and Safety in Industry 5.0
7. A Comprehensive Review of Explainable AI for PHM to Enhancing Trustworthiness in AI-Based Prognostics
8. Human-Robot Collaboration
9. The case studies presented and analyzed in several chapters, as well as, some source codes of algorithms, are also shared with readers.

3 Structure of the Book

The book delves into fundamental principles and recent developments in Artificial Intelligence for Safety and Reliability Engineering, offering in-depth insights across nine chapters.

In the Introductory chapter "Introduction to Artificial Intelligence for Safety and Reliability Engineering" the book editor, K. P. Tran, elaborates on the overview of Artificial Intelligence for Safety and Reliability Engineering.

D. H. Nguyen, T. H. Nguyen, K. D. Tran and K. P. Tran investigate in their chapter, "Physics-Informed Machine Learning for Industrial Reliability and Safety Engineering: A Review and Perspective", a survey on Physics-informed Machine Learning for Industrial Reliability and Safety Engineering to clarify some important issues related to the applications and the potential of Physics-informed Machine Learning algorithms in Industrial Reliability and Safety Engineering.

H. C. Vu, K. D. Tran, V. H. Tran, and K. P. Tran develop in their chapter, "Predictive Maintenance Optimization Based on Genetic Algorithms for Future Industrial Systems", a predictive maintenance approach for future industrial systems that considers the sharing of setup costs, the complexity of system structure, and the system reconfiguration.

M. Orabi, K. P. Tran, S. Thomassey, and P. Egger develop in their chapter, "Anomaly Detection for Catalyzing Operational Excellence in Complex Manufacturing Processes: A Survey and Perspective" the complexities of Smart Manufacturing, proposing advanced AD methodologies designed to cope with the challenges posed by the integration of cutting-edge technologies.

L. H. Nguyen, K. D. Tran, X. Zeng, and K. P. Tran develop in their chapter, "Human-Centered Edge Artificial Intelligence for Smart Factory Applications in Industry 5.0: A Review and Perspective", research perspectives about Explainable AI (XAI), Federated Learning, Self-Supervised Learning, 1-bit machine, and advanced quantization techniques to address these issues about the developing of Human-Centered Edge Artificial Intelligence for Smart Factory Applications in Industry 5.0.

D. C. Nguyen, K. T. P. Nguyen, and K. Medjaher develop in their chapter, "Enhancing Trustworthiness in AI-Based Prognostics: A Comprehensive Review of Explainable AI for PHM" a comprehensive review of XAI in Prognostics and Health Management (PHM), clarifying key concepts, and introducing a structured taxonomy tailored for PHM applications.

M. Benoussaad developed in his chapter, "Human-Robot Collaboration", an overview and state-of-the-art in the field of collaborative robotics, where several issues, challenges, and contributions in the field are summarized. The chapter addresses different aspects of collaborative robotics and its positioning in the context of the whole robotics research.

T. Nguyen, D. H. Nguyen, Q. T. Nguyen, K. D. Tran and K. P. Tran develop in their chapter, "Human-Centered Edge AI and Wearable Technology for Workplace Health and Safety in Industry 5.0" the integration of human-centered edge artificial intelligence (AI) and wearable technology to enhance workplace health and safety within the framework of Industry 5.0.

T. H. Nguyen, A. Saghir, K. D. Tran, D. C. Nguyen, N. A. Luong and K. P. Tran develop in their chapter, "Safety and Reliability of Artificial Intelligence Systems" an overview of AI safety and reliability, discussing major advancements, methodologies for reliability assessment, and practical examples of AI applications.

4 Conclusion

In conclusion, this book introduces readers to the latest advancements in the theoretical applications of Artificial Intelligence (AI) for Safety and Reliability Engineering, as well as ensuring the safety and reliability of AI systems. It addresses AI approaches in response to current challenges and opportunities in Smart Manufacturing within the context of Industry 5.0. Additionally, the book provides ready-to-use algorithms for readers and practitioners to apply AI technologies in Safety and Reliability Engineering and to build trustworthy AI systems.

References

1. Nguyen HD, Tran KP, Zeng X, Koehl L, Castagliola P, Bruniaux P (2019) Industrial internet of things, big data, and artificial intelligence in the smart factory: a survey and perspective. In: ISSAT international conference on data science in business, finance and industry, pp 72–76
2. Huong TT, Bac TP, Long DM, Luong TD, Dan NM, Thang BD, Tran KP et al (2021) Detecting cyberattacks using anomaly detection in industrial control systems: a federated learning approach. Comput Ind 132:103509
3. Du NH, Long NH, Ha KN, Hoang NV, Huong TT, Tran KP (2023) Trans-lighter: a light-weight federated learning-based architecture for remaining useful lifetime prediction. Comput Ind 148:103888
4. Maddikunta PKR, Pham Q-V, Prabadevi B, Deepa N, Dev K, Gadekallu TR, Ruby R, Liyanage M (2022) Industry 5.0: a survey on enabling technologies and potential applications. J Ind Inf Integr 26:100257. ISSN 2452-414X
5. Wang B, Tao F, Fang X, Liu C, Liu Y, Freiheit T (2021) Smart manufacturing and intelligent manufacturing: a comparative review. Engineering 7(6):738–757. ISSN 2095-8099
6. Resek RW (1966) Investment by manufacturing firms: a quarterly time series analysis of industry data. Rev Econ Stat 48(3):322–333. ISSN 00346535, 15309142
7. Zarnowitz V, Ozyildirim A (2006) Time series decomposition and measurement of business cycles, trends and growth cycles. J Monet Econ 53(7):1717–1739. ISSN 0304-3932
8. Hoang NX, Hoang NV, Du NH, Huong TT, Tran KP et al (2022) Explainable anomaly detection for industrial control system cybersecurity. IFAC-PapersOnLine 55(10):1183–1188
9. Truong HT, Ta BP, Le QA, Nguyen DM, Le CT, Nguyen HX, Do HT, Nguyen HT, Tran KP (2022) Light-weight federated learning-based anomaly detection for time-series data in industrial control systems. Comput Ind 140:103692
10. Nguyen HD, Tran KP (2023) Artificial intelligence for smart manufacturing in industry 5.0: Methods, applications, and challenges. Artif Intell Smart Manuf Methods Appl Chall 5–33

Physics-Informed Machine Learning for Industrial Reliability and Safety Engineering: A Review and Perspective

Dac Hieu Nguyen, Thi Hien Nguyen, Kim Duc Tran, and Kim Phuc Tran

Abstract The convergence of physics-informed and machine learning has led to the emergence of Physics-Informed Machine Learning (PIML), a powerful paradigm to enhance the reliability and safety of complex industrial systems. Traditional methods in reliability engineering often rely on physics-based models, which, despite their robustness, face limitations when dealing with highly complex systems involving multiple interacting scales and incomplete knowledge. PIML addresses these challenges by integrating physical laws directly into machine learning models, combining the strengths of data-driven and physics-based approaches. This integration significantly enhances the accuracy and interpretability of predictions, which is crucial for industrial systems' reliability and safety. Inspired by this motivation, this chapter delves into the state-of-the-art developments in PIML, highlighting its applications, benefits, and challenges within reliability and safety engineering.

1 Introduction

Ensuring the long-run operation of complex industrial systems relies heavily on Reliability and Safety Engineering principles. Due to complex series-parallel arrangements in heavy process industries, failure prediction is challenging for system analysts [1]. Moreover, real-time assessment and future state prediction are vital for effective reliability and safety evaluation.

D. H. Nguyen
Institute of Artificial Intelligence and Data Science, Thuyloi University, Hanoi, Vietnam

T. H. Nguyen
Laboratoire AGM, UMR CNRS 8088, CY Cergy Paris Université, 95000 Cergy, France

D. H. Nguyen · K. D. Tran (✉) · K. P. Tran
International Chair in DS & XAI, International Research Institute for Artificial Intelligence and Data Science, Dong A University, Danang, Vietnam
e-mail: ductk@donga.edu.vn

K. P. Tran
ULR 2461 - GEMTEX - Génie et Matériaux Textiles, ENSAIT, University of Lille, 59000 Lille, France

K. P. Tran (ed.), *Artificial Intelligence for Safety and Reliability Engineering*, Springer Series in Reliability Engineering,
https://doi.org/10.1007/978-3-031-71495-5_2

The research outlines the fundamental concepts of PIML, including physics-based models and physics-informed neural networks (PINNs), and explores how these models incorporate physical constraints, such as partial differential equations (PDEs), into the training process to ensure neural network outputs remain consistent with real-world phenomena. Various forward and inverse problems solved using PIML are examined, including predicting temperature fields in additive manufacturing, estimating the remaining useful life of power electronic devices, and modeling fluid dynamics in high-speed boundary layers, demonstrating PIML's ability to provide accurate, real-time assessments and future state predictions. Despite its promising capabilities, PIML faces challenges such as the need for a deep understanding of both machine learning techniques and the underlying physical systems, significant computational resources, and scarcity and noise in training data. The review discusses potential strategies to overcome these hurdles, like developing reduced-order models and integrating advanced neural network architectures. Concluding with a perspective on the future of PIML, the paper emphasizes the need for interdisciplinary collaboration, the development of more efficient algorithms, and the potential for PIML to uncover new physical principles, ultimately revolutionizing how we ensure the reliability and safety of complex industrial systems.

The rest of this chapter is structured as follows: We begin with "An Introduction to Physics-Informed Machine Learning" in Sect. 2. Here, we discuss the pivotal role of integrating physics-based models with machine learning techniques, especially neural networks, to address the limitations of traditional approaches. Subsequently, the chapter overview concept of "Reliability and Safety of Complex Technical Systems and Processes" in Sect. 3, examines the importance of multi-state system reliability in analyzing and optimizing the reliability and safety of complex systems. This is followed by a detailed look at "Physics-Informed Machine Learning for Industrial Reliability and Safety Engineering" in Sect. 4, highlighting various applications of PIML in different industrial sectors, demonstrating its potential to provide accurate, real-time assessments and future state predictions. In "Difficulties, Challenges, and Perspectives for Physics-Informed Machine Learning in Industrial Reliability and Safety Engineering" in Sect. 5, we focus on the challenges and opportunities associated with PIML, emphasizing the need for interdisciplinary expertise and continuous innovation. The chapter continues with a case study in Sect. 6, "Utilize Physics-Informed Wavelet on Fault Diagnosis of Rolling Bearings" showcasing the practical benefits of PIML in fault diagnosis. Finally, Sect. 7 provides perspectives for the future of PIML, highlighting the potential for interdisciplinary collaboration, the development of efficient algorithms, and the discovery of new physical principles.

2 Physics-Informed Machine Learning

In this section, we provide a brief overview of physics-based models and their challenges. It also introduces the integration of physics-based models with machine learning techniques, especially the most popular neural network architectures, to deal with complex systems and mitigate the drawbacks of physics-based approaches.

2.1 *Physics-Based Models*

Almost simulating and predicting the behavior of complex systems using physics-based models involves using mathematical equations and principles derived from physical laws to simulate the behavior of real-world systems. Some of the most popular and widely used physics-based models include Navier–Stokes equations [2] that describe the motion of fluid dynamic, Maxwell's equations [3] describe how electric and magnetic fields interact, Schrödinger Equations [4] describe how the quantum state of a physical system changes over time, etc.

Digital Twins have emerged as a new wave of technology in recent years, and at the core of digital twins are physics-based models. A Digital Twin is a virtual representation of an object or system, created to mirror a physical entity, which can be used to create virtual representations of almost anything, simulating different scenarios and predicting how a system will behave under different conditions, such as minimizing the risk of cyber-attacks and accidents for car [5], prognostics and health management [6], and the supply chain's resilience during the COVID-19 pandemic [7]. According to Research & Markets [8], the global digital twin market is projected to expand from USD 10.1 billion in 2023 to USD 110.1 billion by 2028, achieving a compound annual growth rate of 61.3% during the forecast period.

Physics-based models face significant challenges despite their robustness, particularly when applied to complex systems with multiple interacting scales and incomplete knowledge. Many real-world phenomena' inherent complexity and multi-scale nature require sophisticated models that can accurately capture interactions across different spatial and temporal scales. Additionally, solving the governing equations for large-scale systems often demands substantial computational resources, making high-fidelity simulations both time-consuming and expensive. Several techniques and methods have been developed to address these limitations to enhance the efficiency and accuracy of physics-based models, such as Reduced Order Models, Forward Solving Partial Differential Equations, Inverse Modeling, etc. With the advent of big data and artificial intelligence, several researchers began to use a data-driven approach to complement and enhance traditional physics-based models. The intersection of machine learning and physics allows for integrating physical laws into machine learning models. Conversely, machine learning techniques increasingly enable the discovery of new physical principles that humans haven't been able to discover.

2.2 Physics-Informed Neural Network

PINN was first conceptualized by Raissi et al. [9], who introduced the concept and demonstrated the effectiveness of PINNs in solving forward and inverse problems. The fundamental idea behind PINNs is to approximate the physical system's solution u(t) using a neural network $u_{NN}(t, \theta)$, parameterized by θ [9]

$$u_{NN}(t, \theta) \approx u(t) \tag{1}$$

Traditional neural networks are based purely on a data-driven approach, which does not consider the physical laws. This approach has significant downsides; acquiring a model reliably requires a large amount of data to train, and predictions are almost physically inconsistent and unreliable. An effective way to overcome that limitation is to incorporate physical laws, expressed as PDEs, directly into the loss function of the neural network. In addition to ensuring that the neural network's output satisfies the governing PDE and is consistent with real-world phenomena, it is crucial to incorporate boundary conditions (BCs) and initial conditions (ICs) into the loss function

$$\mathcal{L}(\theta) = \mathcal{L}(\theta)_{data} + \mathcal{L}(\theta)_{PDE} + \mathcal{L}(\theta)_{BC} + \mathcal{L}(\theta)_{IC} \tag{2}$$

where the first term representing the loss originated that measures the difference between the predicted output u_{NN} and the actual output u_{true} is given by

$$\mathcal{L}_{data} = \frac{1}{N_n} \sum_{i=1}^{N_n} (u_{NN}(x_i, t_i, \theta) - u_{true}(x_i, t_i))^2 \tag{3}$$

Depending on the physical problem, selecting appropriate PDEs is crucial for accurate modeling and problem-solving. For instance, to simulate the wavefield [10] or heat transfer [11, 12] uses integer-order PDEs, while fluid mechanics problems [13] often rely on Stochastic PDEs, etc. Consider the 1D heat equation:

$$\frac{\partial u}{\partial t} - k \frac{\partial^2 u}{\partial x^2} = 0 \tag{4}$$

$$\mathcal{L}_{PDE} = \frac{1}{N_p} \sum_{j=1}^{N_p} \left(\left[\frac{\partial u}{\partial t_j} - k \frac{\partial^2 u}{\partial {x_j}^2} \right] u_{NN}(t_j, \theta) \right)^2 \tag{5}$$

BCs and ICs provide necessary constraints that guide the neural network towards physical solutions. Incorporating them into the loss function to optimize the difference between the neural network output and the domain boundaries. The BCs could be Dirichlet, Cauchy, Neumann, Robin, and Mixed boundary conditions. For example, a Dirichlet boundary condition [14–16]:

$$u(x,t) = g(x,t), x \in \delta\Omega, t \in [0,t] \tag{6}$$

$$u(x, t = t_0) = h(x), x \in \Omega \tag{7}$$

$$\mathcal{L}_{BC} = \frac{1}{N_b}\sum_{j=1}^{N_b}\left(u_{NN}(x_j, t_j, \theta) - g(x_j, t_j)\right)^2 \tag{8}$$

$$\mathcal{L}_{IC} = \frac{1}{N_h}\sum_{k=1}^{N_h}\left(u_{NN}(x_j, t_0, \theta) - h(x_k)\right)^2 \tag{9}$$

where $\Omega \in R^d$ with boundaries $\delta\Omega$, N_b and N_i are the number of data points for BCs and ICs.

Incorporating the physics-informed loss function into advanced neural network architectures such as Recurrent Neural Networks (RNNs), Convolutional Neural Networks (CNNs), and Transformer has opened new avenues for modeling dynamic and spatially complex physical systems.

Recurrent Neural Networks

Physics-informed recurrent neural networks (PI-RNNs) have garnered significant attention across various fields due to their capability to integrate domain knowledge and physical laws into neural network models. These networks utilize recurrent neural networks (RNNs) for sequential data processing while incorporating physics-based constraints to enhance prediction accuracy and model physical processes. The mathematical representation of an RNN may be written as follows:

$$h_t = \sigma(W_h h_{t-1} + W_x x_t + b_h) \tag{10}$$

$$y_t = \sigma(W_y h_t + b_y) \tag{11}$$

Where h_t denotes the hidden state at the time, x_t is input at the current time step, W_h and W_x are the weight matrix associated with the prior temporal state, b_h is bias vector, σ is an activation function such as tanh or ReLU. The output y_t at the time t includes W_y as the weight matrix for the output layer and b_y is the bias vector.

The physics-informed loss function typically comprises two parts: the conventional data-driven loss and the physics-based loss [17]. The total loss function $\mathcal{L}$ can be expressed as:

$$\mathcal{L} = w_X \mathcal{L}_X + w_G \mathcal{L}_G \tag{12}$$

where $\mathcal{L}_X$ is the conventional mean squared error (MSE) loss based on training data, $\mathcal{L}_G$ is the physics-based loss incorporating the governing equations, w_X and w_G are weights balancing the contributions of each loss term.

Numerous studies have investigated the application of PI-RNNs in diverse fields such as musculoskeletal systems [18], power electronic devices [19], soft pneumatic actuators [20], energy buffer systems [21] and RRAM modeling [22]. Integrating physics into recurrent neural networks through physics-informed methodologies enhances the interpretability, generalization, and performance of neural network models across different domains, underscoring the potential of PI-RNNs in advancing predictive capabilities while ensuring physical consistency.

Convolutional Neural Networks
Convolutional Neural Networks (CNNs) have shown immense potential in various domains due to their ability to automatically extract hierarchical features from data. Traditional CNN is composed of multiple layers, including convolutional, pooling, and fully connected layers. To incorporate physical constraints, the loss function in a CNN is augmented with a physics-informed term as Eq. 12.

Researchers have explored the use of PINNs in diverse areas such as musculoskeletal modeling [23], turbulence modeling [24], and microstructure-property linkages [25]. These models can effectively capture complex relationships in systems governed by partial differential equations (PDEs) even with limited training data.

Transformer
Transformers have emerged as a leading trend in the field of deep learning in recent years. Self-attention mechanism is the heart of Transformer [26], that the amount of computation can be parallelized, leveraging the power of the GPUs and reducing a huge of time for training and prediction tasks. Lorsung et al. [27] proposed a novel transformer-based architecture, Physics Informed Token Transformer (PITT) designed to learn analytically-driven numerical update operators from governing equations. On a wide variety of challenging 1D and 2D benchmarks, PITT outperforms baseline neural operators and can learn physically significant information from tokenized equations. By embedding physical constraints into the Transformer architecture, this promising approach achieves a more accurate and physically consistent representation of complex systems.

3 Reliability and Safety of Complex Technical Systems and Processes

The reliability and safety of complex technical systems are crucial in ensuring effective operation and preventing failures. These systems are often composed of numerous interdependent components and require sophisticated methods for analysis and optimization. Prioritizing safety and reliability is essential for designing, implementing, and maintaining engineering systems. This section introduces the concept of reliability and safety of complex systems and focuses on the multi-state system reliability approach.

According to the IEEE (Institute of Electrical and Electronics Engineers) standards [28]: "reliability is generally defined as the ability of a system or component to perform its required functions under stated conditions for a specified period". Refer to Kiran et al. [29]: "Reliability engineering is an engineering discipline for applying scientific know-how to a component, product, plant, or process to ensure that it performs its intended function, without failure, for the required time duration in a specified environment". It focuses on dependability throughout a product's lifecycle, emphasizing the ability to endure several years of life under threatening pressures, such as temperature, vibration, and shock. Reliability engineering aims to maintain functional reliability, ensuring quality conformance and customer satisfaction. As products and equipment grow increasingly sophisticated, the demand for higher reliability increases. This area establishes manufacturing schedules and examines how well products function in particular scenarios. It also aims to improve design life, mitigate defects, and reduce failure likelihood. Reliability is mathematically defined as the probability that the random variable representing the time to failure (T) is greater than or equal to the mission time (t), as demonstrated below:

$$R(t) = P(T \geq t)$$

The concept of Safety engineering is related to risk: the higher the risk, the lower the safety, and vice versa. Kaplan and Garrick [30] defined risk by answering the following three questions: "What can go wrong?"; "How likely is it to go wrong?"; "If it does go wrong, what are the consequences?" and calculated in this equation:

$$Risk = \langle S_i P_i x_i \rangle \tag{13}$$

where 'i' denotes a scenario number, S_i denotes an accident scenario with a probability of P_i and a consequence of x_i.

In reliability engineering, a multi-state approach [31], rather than a traditional two-state model, provides a more precise analysis by considering the different states through which the system operation process transitions between the operation states. To identify the degrading components can leverage the multi-state series model, such as "m out of n", consecutive "m out of n", "m out of n"-series, series-consecutive "m out of n" and consecutive "m out of n"-series systems [32]. The multi-state reliability system can be presented:

$$R(t, u) = [R(t, 0), R(t, 1), \ldots, R(t, z)], \quad t \in \langle 0, \infty), \tag{14}$$

where $R(t, u)$, $u = 0, 1, ..., z$ is probability that the system is in the state subset $u, u + 1, ..., z$ at the moment t, $t \in \langle 0, \infty)$, while it was in the state z at the moment $t = 0$.

Researchers have focused on developing models and algorithms to analyze the reliability of multi-state systems under different conditions, such as stress-strength models and dynamic evaluations. Very recently, Gu et al. [33] have proposed an Integrated Availability Importance Measure (IAIM) to evaluate the impact of components on multi-state systems comprehensively. Considering various factors such as component state probability, repair rate, and maintainability, this measure offers a holistic approach to assessing component importance in system reliability.

4 Physics-Informed Machine Learning for Industrial Reliability and Safety Engineering

This section aims to collect cutting-edge research work from eminent researchers in the domain of Physic-informed machine learning (PIML) related to reliability engineering, maintenance decision-making, and risk and safety aspects in different industrial sectors. In the past, the world has faced many industrial disasters due to sudden failure of an equipment/subsystem associated with an industrial system, for example, the most important recent one is the explosion of a stock of ammonium nitrate in a warehouse in Beirut, the capital of Lebanon on August 4, 2020. there were 218 deaths, 7,000 injuries, and a vast number of people homeless (estimated 300,000). An estimated 15 billion was lost on these mishaps overall. Therefore, unswerving truth about improving the reliability and safety of systems is extremely important and urgent.

Industrial Reliability and Safety Engineering is an engineering discipline that assures engineered systems offer satisfactory levels of safety. Reliability and safety have become crucial issues because engineering systems are present in every aspect of social life, including economics, education, healthcare, transportation, food, and scientific research (see Verma et al. [30]). Headformost, we introduce a brief definition of reliability and safety engineering. Recall that the definitions of Reliability and Safety are given by in Sect. 3. While reliability addresses the failure concept, safety addresses the consequences of failure. To put it another way, safety generally refers to ensuring that a failure never happens, whereas reliability generally refers to ensuring that the desired event always happens. Reliability and safety engineering has a wide number of applications in all engineering fields, for example: design evaluation, signal processing, safety management, etc. A great deal of effort was put into developing techniques for enhancing the safety and dependability of systems by numerous academics, practicing engineers, and researchers in a variety of fields. Very recently, the intersection of two different areas of knowledge: physics and machine learning has been strongly developed. The goal is to combine the best aspects of both worlds to more accurate or faster predictions, Physics-Informed Machine Learning (PIML) was born. Physics-Informed Machine Learning (PIML) is the set of techniques that combine physical constraints with statistical machine learning algorithms, including supervised learning, unsupervised learning, semi-supervised learning, reinforcement learning, and etcetera, to solve a scientific issues Hao et al. [34]. PIML is a significant recent development to ensure the reliability and interpretability of systems Farrag et al. [35]. In PIML, we can not ignore physics-informed neural networks (PINNs), and the fact that using the fundamental physical equations to train neural networks. The majority of applications lack sufficient and accurate data, and the physics governing them is unknown, which restricts the applicability of conventional machine learning (ML) and physics-based techniques Ji et al. [36]. Face to this problem, PINNs PINNs are used as a tool to overcome this issue Law [37, 38]. Section 4 in Xu et al. [39] gave a state of the art about PIML for reliability and systems safety applications very comprehensive and detailed. Therefore, in this work, we contribute a look from the other side, starting from the simplest approaches to PIML. PIML can be classified into two main categories, namely, "forward problems" and " inverse problems" Karniadakis et al. [40] and Farrag et al. [35].

4.1 Applications in Forward Problems

Forward problems aim to identify a physical system that satisfies the given data or constraints, it goes from the input to the output. The inputs to the PDEs, such as initial or boundary data, and coefficients, may not always be fully known. Some examples of applications in forward problems: are given boundary conditions and the heat equation, finding the temperature inside the heat sink Farrag et al. [35], predicting molecular properties, and forecasting future weather patterns. Kapoor et al. [41] propose using the PINNs algorithm to solve the forward problems for non-dimensional Euler-Bernoulli systems. The authors conclude that PINNs make it possible to simulate intricate structural systems with numerous interacting components in an accurate, reliable, and efficient manner.

Accelerated Prediction
AI provides the much-needed capability of prediction. Accelerated prediction is extremely important. With applications of PIML, Costabal et al. [42] used a high-resolution cardiac model and multi-fidelity Gaussian process regression to obtain the accelerated prediction of electrocardiogram properties in response to drugs. Another example, in Jiang et al. [43], for the prediction of the temperature and the melt pool dimensions purpose, they used a physics-informed machine learning method. The results demonstrate the high prediction accuracy of the PINN models, which are constrained by the heat transfer equation and certain labeled data.

Uncertainty Quantification
In several real-world scientific applications, uncertainty quantification (UQ) plays an important role because it allows users to assess potential risks related to machine learning (ML) model predictions. A technique called Monte Carlo Dropout (Srivastava et al. [44]) for performing UQ in neural networks is usually used. However, in some cases, this technique is incompatible with the physics-guided learning paradigm. To overcome this limitation, MC Dropout can be combined with a neural network technique known as Physics-guided Architecture (PGA). Therefore, Dropout versions of PINN and APINN (Augmented PINN) were developed (see Zhang et al. [45] and Hu et al. [46]).

Sensitivity Analysis
Sensitivity analyses study how the reliability of the model changes if different sources are changed (Saltelli et al. [47]). This technique is used within specific boundaries that depend on one or more input variables. Physics-informed ML models can outperform basic ML models in terms of prediction accuracy, particularly when there is limited experimental data available. For example, Kapusuzoglu et al. [48] considered two

representative machine learning (ML) models: GP (Gaussian process) and DNN (deep neural networks), incorporating physics knowledge, for sensitivity analysis. The numerical results show that the application of PIML strategies to both these techniques produces physically meaningful results.

4.2 Applications in Inverse Problems

Opposite to the forward problem, the inverse problem goes from the output to the input. Mathematically, the inverse problem can be described as estimating the model's hidden parameters. They are solved by optimizing some functions:

$$\min_{u} D(u, \hat{u}) + R(u)$$

where R is a regularization term that adds some inductive bias for improved generalization, and D is the distance function measuring the discrepancy between the predicted function u and the measurements $\hat{u}$. Numerous fields, including geophysics, mechanical engineering, imaging, and medical imaging, use the inverse problem (Bai et al. [49]). In Cai et al. [50], the authors examined physics-informed fluid mechanics techniques for smoothly integrating data and demonstrating the potency of physics-informed neural networks (PINNs) in inverse problems pertaining to simulating various kinds of flows.

Denoising
Due to inherent qualities, sensor data are often noisy (Gondara [51]). Previous research has suggested techniques for denoising sensor data or modifying the model to account for noisy measurements. Zhang et al. [52] propose a physics-informed denoising method, PILOT (Physics-Informed Learning for denOising Technology) by using an analytical model based on loss functions during training. A DNN algorithm that outperforms earlier algorithms in improving image signal-to-noise was created by Xypakis et al. [53]. In Che and Yin [54], a learning-enhanced Fourier transform profilometry (LeFTP) module and a lightweight DNN were integrated to create a physics-informed deep learning method. Most studies show that the application of PIML produces better results and overcomes the limitations of previous methods. For more information about Physics-informed denoising, see Yin et al. [55], Krishnan et al. [56], Gedon et al. [57], Fathi et al. [58] and Gao et al. [59].

Parameter Identification
Structural-system identification (SI) refers to methods that use data to mark a mathematical model to invert the calculation of structural systems. Structural identification problems in the real world face numerous obstacles. For instance, high dimensionality and complex multivariate partial differential equations (PDEs) make it challenging to train and learn traditional data-driven algorithms. To overcome those obstacles, numerous researchers employed Physic-informed machine learning techniques Guo and Fang [60], Taneja et al. [61], Chen et al. [62], Li and Lee [63], Tondo et al.

[64], Ngo and Lim [65] and Wei et al. [66]. A feature-encoded physics-informed parameter identification neural network (FEPI-PINN), for instance, is proposed by Taneja et al. [61] to simultaneously predict motion and identify parameters of human musculoskeletal systems. The numerical examples demonstrate the effectiveness of the proposed framework in identifying muscle parameters and resolving the high dimensionality issue.

5 Difficulties, Challenges, and Perspectives for Physics-Informed Machine Learning in Industrial Reliability and Safety Engineering

This section considers the potential challenges and opportunities of PIML methods from various perspectives. We can not deny that physics-informed Machine Learning improves the quality and reconstruction time for certain models because it is a promising solution in the case of sparse data and incomplete physics knowledge. As previously mentioned, significant advancements in industrial reliability and safety engineering have been made in PIML in recent years. However, besides that, those models also have some limitations.

5.1 Limitations

Each different PIML model has its limitations. For instance, [67] used DeepONet (Deep operator neural networks) to predict instability waves in high-speed boundary layers. This approach is entirely data-driven, and the amount of training data needed is typically rather large, particularly for some complex PDEs. PINNs (Physics-Informed Neural Networks) are the first neural network architecture to solve forward and inverse PDE problems by incorporating physical knowledge. While this framework is succinct and adaptable, it has numerous drawbacks. For example, different loss terms in PINNs might have very different convergence speeds (Hao et al. [34]). Some limitations of PIML in model properties like: optimization process, prediction results, or model parameter have been reviewed very carefully in Xu et al. [39]. In this work, we emphasize some limitations/difficulties of PILM that most researchers or producers using PIML for their systems often encounter:

Construction Model

The construction of PIML models necessitates a deep understanding of the system's physical properties. For this, the researcher needs to be very knowledgeable about mathematical physics and computer science (e.g., EDP, mechanics) (Hao et al. [34]). Sometimes the models can be too complex to be coded in a machine learning language. As a result, creating a knowledge foundation for PIML from ambiguous data is crucial.

Costs Associated with Gathering Data
Experiments or numerical approaches are used to generate the samples, and both have prohibitive costs and multiple sources of uncertainty, particularly when we face solving high dimensional problems (Hu et al. [68]). How to optimize this cost is the main purpose of the models (see, for example, Alenezi et al. [69], Li et al. [70] and Zhang et al. [71]).

5.2 *Challenges and Perspectives of PIML*

PIML is a very young and fast-growing field, where many open challenges remain to be solved (Hao et al. [34]). Other reviews have emphasized the difficulties with sparse and noisy data, data availability, and incomplete physics models (see Xu et al. [39] and Willard et al. [72]). In this work, we focus on the difficulties of transforming different types of knowledge into the kind required by the ML framework, i.e., fitting knowledge to ML models instead of choosing ML models to fit physics knowledge. As a result, the difficulties and viewpoints can be emphasized in three ways:

Various Branches of Physics and Machine Learning Algorithms
There are still a lot of problems in the field of computational physics to be explored. For example, a promising approach to solving complex computational problems is the intersection of machine learning and quantum computing. Particle and nuclear physics research advances our understanding of the cosmos and pushes accuracy bounds. In this area, the use of artificial intelligence (AI) and machine learning algorithms helps us advance research more quickly and produce outcomes that are beyond our wildest dreams. Fighting global issues, particularly natural disasters, is a more practical field. Machine learning for fluid systems and dynamics has been used extensively in recent years, but improving the already excellent results is still a significant challenge. Some other branches of physics that are still waiting for the applications of ML algorithms to explode can be mentioned as biophysics, thermodynamics, and nanophysics. Some new ML technologies can be applied, for instance, quantum computing and analog computing.

Synthesis of Explicit Model Predictive Control
An indisputable truth is that machine learning and artificial intelligence algorithms are developing at an astonishing rate every day. Selecting and combining the benefits of each algorithm is a perpetually evolving field of study and research.

Providing Safety and Performance Guarantees
No matter how complicated the methods are, the ultimate goal is to optimize the following three factors: safety, reliability, and low costs. Considering limitations in PIML as challenges to overcome is the objective of future researchers in this field. A few questions that naturally come up for researchers/model applicators when they use ML algorithms in their model are: How can representative training data be chosen effectively? How can modeling errors and uncertainty be quantified for PIML-based

models? How can the computational demands of high-fidelity digital twins be met without compromising accuracy? How to ensure a real-world system's safety and stability.

6 Case Study: Utilize Physics-Informed Wavelet on Fault Diagnosis of Rolling Bearings

This section presents an implementation of PINN for intelligent fault diagnosis of rolling bearings. Rolling bearings are vital in rotating machinery, and their failure can lead to significant accidents, making intelligent fault diagnosis with deep learning technology, enhanced by advances in AI and computational power, critically important. Very recently, to improve training speed and model cooperation, Chao He et al. [73] have introduced a novel and general deep triple-stream network called EWSNet. It is incorporated in the threshold noise reduction algorithm and wavelet signal processing approach. In this case study, we will utilize EWSNet to demonstrate the effectiveness of integrating physical prior knowledge into deep neural networks.

EWSNet enhanced wavelet basis function on the first convolution layer based on CNN, it is called enhanced wavelet convolution (EWconv). In the original research paper, the proposed wavelet weight initialization incorporates five wavelet basis functions, including Laplace, Morlet, Mexican Hat, Gaussian, and Shannon. This experiment will use Laplace, the equation of the Laplace wavelet basis function is as follows:

$$\psi(t) = Ae^{\frac{-\xi}{\sqrt{1-\xi^2}} \times 2\pi f(t-\tau)} \times \sin[2\pi f(t-\tau)] \tag{15}$$

where f denotes the sampling frequency, ξ denotes the viscous damping ratio, A denotes a wavelet normalization function, τ denotes a time parameter.

In order to improve the interpretability and trainability of the wavelet transform, EWSNet additionally defined a scale smoothing factor to maintain consistency between the translation and scale factors, the equation of the scale smoothing factor ζ is as follows:

$$\psi_{u,s}^{L}(t) = Ae^{\frac{-\xi}{\sqrt{1-\xi^2}} \times 2\pi f\left(\frac{t-u}{s-\zeta}-\tau\right)} \times \sin\left[2\pi f\left(\frac{t-u}{s-\zeta}-\tau\right)\right] \tag{16}$$

In the experimental benchmarks on the JNU data set, bearings are classified into four states: normal (NC), inner race fault (IF), outer race fault (OF), and ball fault (BF). We examined the speed case of 800 r/m, the sample length is 1024 with a sliding window, and then used the Z-index to normalize it. The structure of EWSNet is the same as the original research paper, available at: https://github.com/liguge/EWSNet. The experimental results compared vanilla CNN (vCNN) and EWSNet, which is initialized by wavelet weights to the first convolution layer of vCNN. The

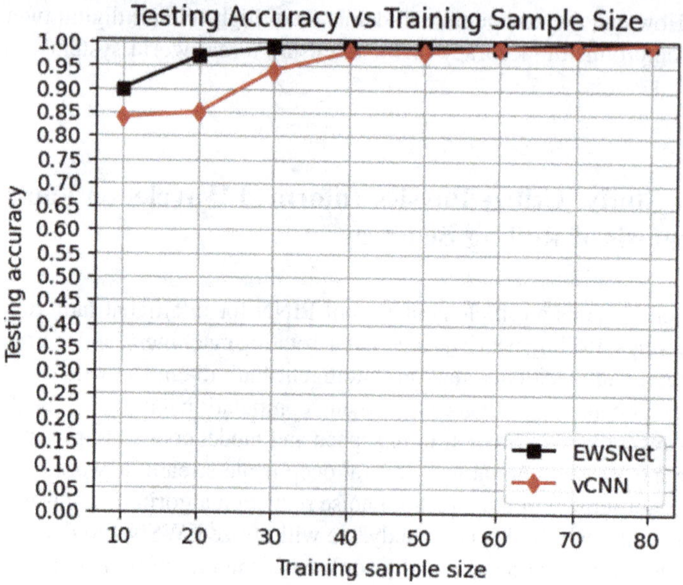

Fig. 1 The accuracy of vCNN and EWSNet with training samples are from 10 to 80

defined parameters are as follows: batch size is 16, $\mu = 0.001$ and $\tau = 0.1$. Each state retrieves from 10 to 80 training samples with step 10. The experimental results are shown in Fig. 1.

The experimental results are almost similar to the original research paper. With 10 random training samples per state, vCNN achieves approximately 84%, while EWSNet has much higher accuracy than vCNN; it archives approximately 90%. Demonstrating that vCNN are embedded wavelet weight initialization with generics and physical interpretability significantly enhances the model's ability to identify fault classification under various conditions, even with limited training sample data.

7 Concluding Remarks

In conclusion, integrating physics-informed machine learning (PIML) into industrial reliability and safety engineering represents a transformative shift in addressing the complexities of modern systems. This review has underscored the potential of PIML to bridge the gap between traditional physics-based models and contemporary data-driven approaches. By embedding physical laws directly into machine learning frameworks, PIML enhances predictions' accuracy, interpretability, and reliability, which is crucial for the safe and efficient operation of complex industrial systems.

The applications of PIML are vast, from predicting temperature fields in additive manufacturing to modeling fluid dynamics and estimating the remaining useful life of power electronic devices. These examples demonstrate the ability of PIML to provide real-time assessments and predictive insights consistent with physical reality, thus enabling proactive maintenance and risk management strategies.

However, the adoption of PIML is not without challenges. The primary hurdles are the need for interdisciplinary expertise, significant computational resources, and high-quality training data. Moreover, developing efficient algorithms and integrating advanced neural network architectures remain critical areas for future research. Addressing these challenges requires ongoing collaboration across disciplines, continuous innovation in algorithm development, and the exploration of new applications to realize the full potential of PIML.

Looking forward, the future of PIML in industrial reliability and safety engineering is promising. Continued advancements in this field hold the potential to uncover new physical principles and revolutionize the way we approach the design, operation, and maintenance of complex systems. Emphasizing the need for interdisciplinary collaboration and developing more efficient and robust algorithms will be key to overcoming current limitations and driving further innovations.

In summary, PIML offers a powerful framework for enhancing the reliability and safety of industrial systems. Its ability to integrate physical constraints with machine learning opens new avenues for predictive maintenance, risk management, and system optimization. As research in this area progresses, PIML is poised to play a pivotal role in ensuring the safe and reliable operation of industrial systems in the future.

References

1. Panchal D, Ram M, Chatterjee P, Sachdeva AK (2023) Applications and practices. Industrial reliability and safety engineering. CRC Press
2. Heywood JG (2006) Viscous incompressible fluids: mathematical theory. In: Françoise J-P, Naber GL, Tsun TS (eds), Encyclopedia of mathematical physics, pp 369–379. Academic, Oxford. ISBN 978-0-12-512666-3. https://doi.org/10.1016/B0-12-512666-2/00158-9. https://www.sciencedirect.com/science/article/pii/B0125126662001589
3. Fleisch D (2008) A student's guide to Maxwell's equations. Cambridge University Press
4. Berezin FA, Shubin M (2012) The Schrödinger equation, vol 66. Springer Science & Business Media
5. Almeaibed S, Al-Rubaye S, Tsourdos A, Avdelidis NP (2021) Digital twin analysis to promote safety and security in autonomous vehicles. IEEE Commun Stand Mag 5(1):40–46. https://doi.org/10.1109/MCOMSTD.011.2100004
6. Tao F, Zhang M, Liu Y, Nee AYC (2018) Digital twin driven prognostics and health management for complex equipment. CIRP Ann 67(1):169–172. ISSN 0007-8506. https://doi.org/10.1016/j.cirp.2018.04.055. https://www.sciencedirect.com/science/article/pii/S0007850618300799
7. Lv Z, Qiao L, Mardani A, Lv H (2022) Digital twins on the resilience of supply chain under covid-19 pandemic. IEEE Trans Eng Manag 1–12. https://doi.org/10.1109/TEM.2022.3195903
8. Research ltd and Markets. Global digital twin market by application (predictive maintenance, business optimization, performance monitoring, inventory management), industry (automotive & transportation, healthcare, energy & utilities), enterprise and geography - forecast to

2028. https://www.researchandmarkets.com/reports/5146336/global-digital-twin-market-by-application
9. Raissi M, Perdikaris P, Karniadakis GE (2019) Physics-informed neural networks: a deep learning framework for solving forward and inverse problems involving nonlinear partial differential equations. J Comput Phys 378:686–707. ISSN 0021-9991. https://doi.org/10.1016/j.jcp.2018.10.045. https://www.sciencedirect.com/science/article/pii/S0021999118307125
10. Moseley B, Markham A, Nissen-Meyer T (2020) Solving the wave equation with physics-informed deep learning. arXiv: Computational Physics. https://api.semanticscholar.org/CorpusID:219966048
11. Zobeiry N, Humfeld KD (2021) A physics-informed machine learning approach for solving heat transfer equation in advanced manufacturing and engineering applications. Eng Appl Artif Intell 101:104232. ISSN 0952-1976. https://doi.org/10.1016/j.engappai.2021.104232. https://www.sciencedirect.com/science/article/pii/S0952197621000798
12. Cai S, Wang Z, Wang S, Perdikaris P, Karniadakis GE (2021) Physics-informed neural networks for heat transfer problems. J Heat Transfer 143(6):060801. ISSN 0022-1481. https://doi.org/10.1115/1.4050542
13. Cai S, Mao Z, Wang Z, Yin M, Karniadakis GE (2021) Physics-informed neural networks (pinns) for fluid mechanics: a review. Acta Mech Sinica 37(12):1727–1738. ISSN 1614-3116. https://doi.org/10.1007/s10409-021-01148-1
14. Berrone S, Canuto C, Pintore M, Sukumar N (2023) Enforcing Dirichlet boundary conditions in physics-informed neural networks and variational physics-informed neural networks. Heliyon 9(8):e18820
15. Wang J, Mo YL, Izzuddin B, Kim C-W (2023) Exact dirichlet boundary physics-informed neural network epinn for solid mechanics. Comput Methods Appl Mech Eng 414:116184. ISSN 0045-7825. https://doi.org/10.1016/j.cma.2023.116184. https://www.sciencedirect.com/science/article/pii/S0045782523003080
16. Hsieh T-Y, Huang T-H (2024) A multiscale stabilized physics informed neural networks with weakly imposed boundary conditions transfer learning method for modeling advection dominated flow. Eng Comput. ISSN 1435-5663. https://doi.org/10.1007/s00366-024-01981-5
17. Zheng Y, Hu C, Wang X, Wu Z (2023) Physics-informed recurrent neural network modeling for predictive control of nonlinear processes. J Process Control 128:103005. ISSN 0959-1524. https://doi.org/10.1016/j.jprocont.2023.103005. https://www.sciencedirect.com/science/article/pii/S0959152423000847
18. Taneja K, He X, He Q, Chen J-S (2023) A multi-resolution physics-informed recurrent neural network: formulation and application to musculoskeletal systems. Comput Mech 73(5):1125–1145. ISSN 1432-0924. https://doi.org/10.1007/s00466-023-02403-x. http://dx.doi.org/10.1007/s00466-023-02403-x
19. Lu Z, Guo C, Liu M, Shi R (2023) Remaining useful lifetime estimation for discrete power electronic devices using physics-informed neural network. Sci Rep 13(1). ISSN 2045-2322. https://doi.org/10.1038/s41598-023-37154-5. http://dx.doi.org/10.1038/s41598-023-37154-5
20. Sun W, Akashi N, Kuniyoshi Y, Nakajima K (2022) Physics-informed recurrent neural networks for soft pneumatic actuators. IEEE Robot Autom Lett 7(3):6862–6869. ISSN 2377-3774. https://doi.org/10.1109/lra.2022.3178496. http://dx.doi.org/10.1109/LRA.2022.3178496
21. Lahariya M, Karami F, Develder C, Crevecoeur G (2021) Physics-informed recurrent neural networks for the identification of a generic energy buffer system. In: 2021 IEEE 10th data driven control and learning systems conference (DDCLS). IEEE
22. Sha Y, Lan J, Li Y, Chen Q (2023) A physics-informed recurrent neural network for rram modeling. In: 2003 international symposium of electronics design automation (ISEDA). IEEE
23. Zhang J, Zhao Y, Shone F, Li Z, Frangi AF, Xie SQ, Zhang Z-Q (2023) Physics-informed deep learning for musculoskeletal modeling: predicting muscle forces and joint kinematics from surface emg. IEEE Trans Neural Syst Rehab Eng 31:484–493. https://doi.org/10.1109/TNSRE.2022.3226860

24. Karpov PI, Huang C, Sitdikov I, Fryer CL, Woosley S, Pilania G (2022) Physics-informed machine learning for modeling turbulence in supernovae. Astrophys J 940(1):26. ISSN 1538-4357. https://doi.org/10.3847/1538-4357/ac88cc. http://dx.doi.org/10.3847/1538-4357/ac88cc
25. Mann A, Kalidindi SR (2022) Development of a robust cnn model for capturing microstructure-property linkages and building property closures supporting material design. Frontiers in Materials. ISSN 2296-8016. https://doi.org/10.3389/fmats.2022.851085. http://dx.doi.org/10.3389/fmats.2022.851085
26. Vaswani A, Shazeer N, Parmar N, Uszkoreit J, Jones L, Gomez AN, Kaiser Ł, Polosukhin I (2017) Attention is all you need. In: Guyon I, Von Luxburg U, Bengio S, Wallach H, Fergus R, Vishwanathan S, Garnett R (eds), Advances in neural information processing systems, vol 30. Curran Associates, Inc. https://proceedings.neurips.cc/paper_files/paper/2017/file/3f5ee243547dee91fbd053c1c4a845aa-Paper.pdf
27. Lorsung C, Li Z, Barati Farimani A (2024) Physics informed token transformer for solving partial differential equations. Mach Learn: Sci Technol 5(1):015032. ISSN 2632-2153. https://doi.org/10.1088/2632-2153/ad27e3. http://dx.doi.org/10.1088/2632-2153/ad27e3
28. IEEE standard computer dictionary (1991) a compilation of ieee standard computer glossaries. IEEE Std 610:1–217. https://doi.org/10.1109/IEEESTD.1991.106963
29. Kiran DR (2017) Reliability engineering (Chap 27). In: Kiran DR (ed), Total quality management, pp 391–404. Butterworth-Heinemann. ISBN 978-0-12-811035-1. https://doi.org/10.1016/B978-0-12-811035-5.00027-1. https://www.sciencedirect.com/science/article/pii/B9780128110355000271
30. Verma A, Srividya A, Karanki DR (2016) Reliability and safety engineering, 2nd edn. ISBN 978-1-4471-6268-1
31. Kołowrocki K, Soszyńska-Budny J (2011) Reliability and safety of complex technical systems and processes: modeling - identification - prediction - optimization. Springer, London
32. Soszynska-Budny J (2011). Safety and reliability of complex technical systems. https://doi.org/10.1109/ICQR2MSE.2011.5976558
33. Dongwei GG, Yuhong Z, Yanjuan H, Guang C, Zhixin W, Nianhuan L (2023) Integrated availability importance measure for multi-state complex systems analysis. Proc Inst Mech Eng Part O: J Risk Reliab 238(3):578–590. https://doi.org/10.1177/1748006X231159823
34. Hao Z, Liu S, Zhang Y, Ying C, Feng Y, Su H, Zhu J (2022) Physics-informed machine learning: a survey on problems, methods and applications
35. Farrag A, Yang Y, Cao N, Won D, Jin Y (2024) Physics-informed machine learning for metal additive manufacturing. Progr Addit Manuf. ISSN 2363-9512. https://doi.org/10.1007/s40964-024-00612-1
36. Ji W, Qiu W, Shi Z, Pan S, Deng S (2021) Stiff-pinn: physics-informed neural network for stiff chemical kinetics. J Phys Chem A 125:8098–8106
37. Lawal ZK, Yassin H, Lai DTC, Idris AC (2022) Physics-informed neural network (pinn) evolution and beyond: a systematic literature review and bibliometric analysis. Big Data Cognit Comput 6(4):140
38. Pateras J, Rana P, Ghosh P (2022) A taxonomic survey of physics-informed machine learning. Appl Sci-Basel 13(12). https://doi.org/10.3390/app13126892
39. Xu Y, Kohtz S, Boakye J, Gardoni P, Wang P (2023) Physics-informed machine learning for reliability and systems safety applications: state of the art and challenges. Reliab Eng & Syst Saf 230:108900. ISSN 0951-8320
40. Karniadakis GE, Kevrekidis IG, Lu L, Perdikaris P, Wang S, Yang L (2021) Physics-informed machine learning. Nat Rev Phys 3
41. Kapoor T, Wang H, Núñez A, Dollevoet R (2023) Physics-informed neural networks for solving forward and inverse problems in complex beam systems. IEEE Trans Neural Netw Learn Syst 1–15:09
42. Costabal F, Matsuno K, Yao J, Perdikaris P, Kuhl E (2019) Machine learning in drug development: characterizing the effect of 30 drugs on the qt interval using gaussian process regression, sensitivity analysis, and uncertainty quantification. Comput Methods Appl Mech Eng 348:02

43. Jiang F, Xia M, Hu Y (2023) Physics-informed machine learning for accurate prediction of temperature and melt pool dimension in metal additive manufacturing. 3D Print Addit Manuf. https://doi.org/10.1089/3dp.2022.0363
44. Srivastava N, Hinton G, Krizhevsky A, Sutskever I, Salakhutdinov R (2014) Dropout: a simple way to prevent neural networks from overfitting. J Mach Learn Res 15:1929–1958
45. Zhang D, Lu L, Guo L, Karniadakis G (2018) Quantifying total uncertainty in physics-informed neural networks for solving forward and inverse stochastic problems
46. Hu Z, Jagtap AD, Karniadakis GE, Kawaguchi K (2023) Augmented physics-informed neural networks (apinns): a gating network-based soft domain decomposition methodology. Eng Appl Artif Intell 126:107183. ISSN 0952-1976
47. Saltelli A, Tarantola S, Chan K (2001) A quantitative model-independent method for global sensitivity analysis of model output. Technometrics 41. https://doi.org/10.1080/00401706.1999.10485594
48. Kapusuzoglu B, Mahadevan S (2021) Information fusion and machine learning for sensitivity analysis using physics knowledge and experimental data. Reliab Eng & Syst Saf 214:107712. ISSN 0951-8320
49. Bai Y, Chen W, Chen J, Guo W (2020) Deep learning methods for solving linear inverse problems: research directions and paradigms. Signal Proc 177:107729. ISSN 0165-1684
50. Cai S, Mao Z, Wang Z, Yin M, Karniadakis GE (2022) Physics-informed neural networks (pinns) for fluid mechanics: a review
51. Gondara L (2016) Medical image denoising using convolutional denoising autoencoders. In: 2016 IEEE 16th international conference on data mining workshops (ICDMW), pp 241–246. https://doi.org/10.1109/ICDMW.2016.0041
52. Zhang X, Fu X, Teng D, Dong C, Vijayakumar K, Zhang J, Chowdhury RR, Han J, Hong D, Kulkarni R, Shang J, Gupta R (2024) Physics-informed data denoising for real-life sensing systems, pp 83–96. https://doi.org/10.1145/3625687.3625811
53. Xypakis E, De Turris V, Gala F, Ruocco G, Leonetti M (2023) Physics-informed deep neural network for image denoising. Opt Express 31(26):43838–43849
54. Che Y, Yin W (2023) A lightweight fringe analysis network based on deep learning. In: Shao X, Cao L, Cao X (eds), AOPC 2023: computing imaging technology. Proceedings of SPIE, vol 12967. Conference on AOPC - computing imaging technology, Beijing, Peoples R China, Jul 25–27
55. Yin W, Che Y, Li X, Li M, Hu Y, Feng S, Lam EY, Chen Q, Zuo C (2024) Physics-informed deep learning for fringe pattern analysis. Opto-Electr Adv 7(1)
56. Krishnan G, Goswami S, Joshi R, Javidi B (2024) Three-dimensional integral imaging-based image descattering and recovery using physics informed unsupervised cyclegan. Optics Express 32(2):1825–1835
57. Gedon D, Ribeiro AH, Wahlstrom N, Schon TB (2023) Invertible kernel pca with random fourier features. IEEE Signal Proc Lett 30:563–567
58. Fathi MF, Perez-Raya I, Baghaie A, Berg P, Janiga G, Arzani A, D'Souza RM (2020) Super-resolution and denoising of 4d-flow mri using physics-informed deep neural nets. Comput Methods Progr Biomed 197
59. Gao H, Sun L, Wang J-X (2021) Super-resolution and denoising of fluid flow using physics-informed convolutional neural networks without high-resolution labels. Phys Fluids 33(7)
60. Guo X-Y, Fang S-E (2023) Structural parameter identification using physics-informed neural networks. Measurement 220
61. Taneja K, He X, He Q, Zhao X, Lin Y-A, Loh KJ, Chen J-S (2022) A feature-encoded physics-informed parameter identification neural network for musculoskeletal systems. J Biomech Eng-Trans Asme 144(12)
62. Chen S, Zhang J, Wang S, Wen P, Zhao S (2022) Circuit parameter identification of degrading dc-dc converters based on physics-informed neural network. In: Li C, Valentino G, Kang L, Cabrera D, Gjorgjevikj D (eds), 2022 prognostics and health management conference, Phm-London 2022, pp 260–268. Prognostics and health management conference (PHM-2022 London). London, England

63. Li W, Lee K-M (2021) Physics informed neural network for parameter identification and boundary force estimation of compliant and biomechanical systems. Int J Intell Robot Appl 5(3, SI):313–325
64. Tondo GR, Rau S, Kavrakov I, Morgenthal G (2023) Stochastic stiffness identification and response estimation of timoshenko beams via physics-informed gaussian processes. Probab Eng Mech 74
65. Ngo SI, Lim Y-I (2021) Solution and parameter identification of a fixed-bed reactor model for catalytic co_2 methanation using physics-informed neural networks. Catalysts 11(11)
66. Wei Y, Serra Q, Lubineau G, Florentin E (2023) Coupling physics-informed neural networks and constitutive relation error concept to solve a parameter identification problem. Comput & Struct 283
67. di Leoni PC, Meneveau C, Karniadakis G, Zaki T (2020) Deep Operator Neural Networks (DeepONets) for prediction of instability waves in high-speed boundary layers. In: APS division of fluid dynamics meeting abstracts, APS Meeting Abstracts, p R01.004
68. Hu Z, Shi Z, Karniadakis GE, Kawaguchi K (2024) Hutchinson trace estimation for high-dimensional and high-order physics-informed neural networks. Comput Methods Appl Mech Eng 424
69. Alenezi DF, Biehler M, Shi J, Li J (2024) Physics-informed weakly-supervised learning for quality prediction of manufacturing processes. IEEE Trans Autom Sci Eng
70. Li H, Zhang Z, Li T, Si X (2024) A review on physics-informed data-driven remaining useful life prediction: challenges and opportunities. Mech Syst Signal Proc 209
71. Zhang J, Ruan Z, Li Q, Zhang Z-Q (2023) Toward robust and efficient musculoskeletal modeling using distributed physics-informed deep learning. IEEE Trans Instrum Meas 72
72. Willard J, Jia X, Xu S, Steinbach M, Kumar V (2020) Integrating physics-based modeling with machine learning: a survey
73. He C, Shi H, Si J, Li J (2023) Physics-informed interpretable wavelet weight initialization and balanced dynamic adaptive threshold for intelligent fault diagnosis of rolling bearings. J Manuf Syst 70:579–592. ISSN 0278-6125. https://doi.org/10.1016/j.jmsy.2023.08.014. https://www.sciencedirect.com/science/article/pii/S0278612523001644

Predictive Maintenance Optimization Based on Genetic Algorithms for Future Industrial Systems

Hai-Canh Vu, Kim Duc Tran, Viet Hieu Tran, and Kim Phuc Tran

Abstract Predictive maintenance (PdM) is a crucial technology for the industry's future. It involves making maintenance decisions based on the prediction of the system's performance in the future. It helps to reduce maintenance costs and ensures operational efficiency and flexibility of the future industrial systems (FIS) under different dynamic and uncertain conditions. However, applying traditional PdM optimization approaches to FIS is challenging due to the complex interactions and interconnections among the FIS components and the uncertainties of future events. This chapter presents a promising PdM optimization approach based on Genetic Algorithms (GA) for FIS that considers all the above challenges. Then, we illustrate the application of this approach for maintenance optimization of a 16-component system in both normal and dynamic situations where the system structure is changed due to reconfiguration.

Keywords Predictive maintenance · Maintenance optimization · Genetic algorithms · Future industry

---

H.-C. Vu
Roberval Laboratory, University of Technologies of Compiègne, Compiègne, France

K. D. Tran (✉) · V. H. Tran · K. P. Tran
International Chair in DS & XAI, International Research Institute for Artificial Intelligence and Data Science, Dong A University, Danang, Vietnam
e-mail: ductk@donga.edu.vn

K. P. Tran
ULR 2461 - GEMTEX - Génie et Matériaux Textiles, ENSAIT, University of Lille, 59000 Lille, France

V. H. Tran
Faculty of Information Technology, Dong A University, Danang, Vietnam

K. P. Tran (ed.), *Artificial Intelligence for Safety and Reliability Engineering*, Springer Series in Reliability Engineering,
https://doi.org/10.1007/978-3-031-71495-5_3

1 Introduction

Nowadays, future industry has gained much attention from both industry and academia. This includes organizations such as the German Federal Ministry of Education and Research (BMBF), the European Commission, Siemens, Bosch Companies, and GOSPI. It creates what has been called "Future Industrial Systems (FIS)". Within the modular structured smart systems, Cyber-Physical Modules (CPM—integrations of computation, networking, and physical processes) monitor physical processes, create a virtual copy of the physical world, and make decentralized decisions. Over the Internet of Things (network of physical devices), CPMs communicate and cooperate and with humans in real-time, and via cloud computing, both internal and cross-organizational services are offered and used by participants of the value chain [1]. Thanks to the decentralized decision-making and communication between CPMs, the FIS (flexible manufacturing systems) can react proactively to unexpected events, such as dynamic changes in production and component failures. This enables them to operate flexibly and efficiently. For instance, a cyber-physical module can autonomously diagnose and predict its failures based on various data sources. It can then trigger the required maintenance at the best possible time to achieve just-in-time maintenance, ensuring near-zero downtime. The FIS is also able to proactively organize its production and logistic supports in response to the predicted changes in production.

In this context of the future industry, predictive maintenance (PdM) plays a crucial role in ensuring the FIS operational efficiency and flexibility under different dynamic and uncertain conditions [2]. The implementation of the PdM consists of three main steps: data acquisition and preprocessing, prognostics, and maintenance optimization (Fig. 1). Data acquisition and preprocessing are crucial in providing accurate data representing a system or component's health or degradation state. In practice, different health indicators (HIs), such as reliability (safety-critical systems), Remaining

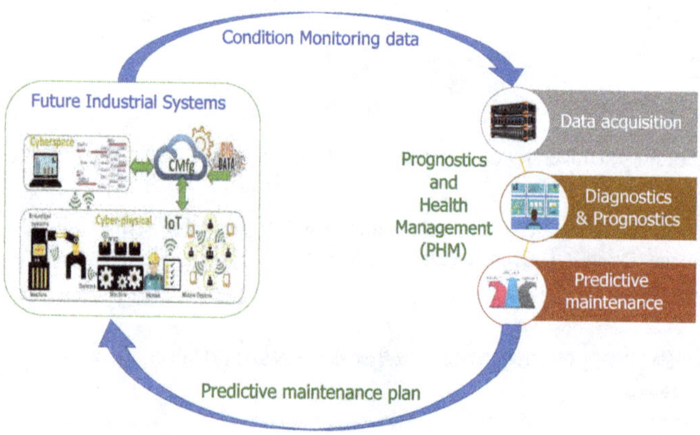

Fig. 1 Predictive maintenance for future industrial systems

Useful Life (RUL), and production rate (production systems), are typically used to quantify health and performance. Prognostics are performed to predict the evolution of HIs based on data, system knowledge, and future mission profiles. The last step, maintenance optimization, involves making the optimal maintenance decisions according to the predicted values of HIs. Unfortunately, given all the advantages of the PdM, it has been mainly developed for traditional industrial systems. This vision is not sufficient for the future industrial systems. Applying traditional PdM models to the FIS is challenging due to complex interactions and interconnections among the FIS components, and uncertainties and dynamism of future related events. In this chapter, we focus on maintenance optimization (the last step of the PdM), and present a promising model based on Genetic Algorithms (GA) for the FIS, considering all the above challenges.

2 Background and Related Work

For the past few decades, there has been an increasing focus on maintenance optimization as it plays a crucial role in reducing maintenance costs, improving system reliability, maintainability, safety, and sustainability [3, 4]. There are two types of maintenance optimization models: mono-component models and multi-component models [5]. Mono-component models focus on optimizing the maintenance of a single component, while multi-component models consider the maintenance of several components by taking into account their dependencies [6]. There are various types of dependencies, including economic, stochastic, and resource dependencies, as discussed in Keizer's work [7]. Economic dependence is the most widely researched in the literature [8]. It suggests that when a group of components is maintained together, their total maintenance cost may change. The phenomenon of economic dependence is deemed positive when joint maintenance is more cost-effective than individual or separate maintenance due to shared tasks, such as machine opening and spare part transportation, as well as system downtimes between maintenance activities. However, it can also have adverse effects, such as component useful life loss, difficulties in maintenance support, and an increase in system failure frequency. While economic dependence can make multi-component models more intricate, it also presents opportunities for cost savings in maintenance if the positive impact is considered. There are two types of maintenance optimization models for multi-component systems that serve this purpose: the grouping maintenance model and the opportunistic maintenance model.

The opportunistic maintenance model is a comprehensive maintenance strategy that encompasses a wide range of maintenance opportunities, including both preventive maintenance and failure of component/system [9–11]. To seize these opportunities, it is typical to advance the preventive maintenance of components. While the opportunistic maintenance model can exploit system downtimes and preparation tasks, it may also result in a loss of component useful life. The maintenance plan established by this model may comprise a combination of both corrective and preventive maintenance activities; however, since corrective maintenance and component failures are unpredictable, the maintenance management and support of the opportunistic maintenance model are usually complex and not suitable for many real applications.

The grouping maintenance model suggests grouping or jointly executing preventive maintenance (PM) activities of multiple components to take advantage of the positive economic dependence between them. This leads to savings in support costs paid for the different maintenance preparation tasks. The maintenance plan provided by this model contains only PM activities and is more reasonable for maintenance management and support. This model has recently been extensively developed in both industry and academia [12–14]. The maintenance grouping models can be classified based on their flexibility level: stationary grouping models and dynamic grouping models. The stationary models offer fixed guidelines for maintenance, which remain constant over time. These guidelines can be based on maintenance frequencies for related activities [15–17] or control thresholds for maintenance actions (e.g., age, reliability, failure rate, remaining useful life thresholds) [18–22]. These models may not be suitable for FIS that operate under dynamic conditions such as varying component deterioration, unpredictable maintenance opportunities, changes in utilization factors, system reconfiguration, etc.

To establish a flexible maintenance plan that can be updated to take advantage of the dynamic conditions, dynamic grouping models that utilize a rolling horizon approach were first introduced in 1997 by Wildeman et al. [23]. The maintenance models consist of three main steps: *(1)* the first step involves determining the maintenance frequency for each component separately on an infinite horizon; *(2)* the second step is to consider a finite horizon and establish an individual maintenance plan based on the maintenance frequency of each component; *(3)* In the third step, dynamic programming optimally groups individual maintenance dates within a finite horizon to save the maintenance costs. This process is then repeated for the other finite planning horizons. The dynamic grouping models allow for the maintenance plan to be updated while ensuring the grouping's short- and long-term performance. They have garnered significant attention and have been extensively developed in numerous works [24–30]. For instance, Bouvard et al. [24] developed a rolling horizon approach for implementing condition-based maintenance on commercial heavy vehicles. This approach was carefully crafted to incorporate maintenance durations and time-limited opportunities [25], as well as the availability constraint under limited repairmen [26]. It is important to clarify that these papers are exclusively focused on series structures that exhibit positive economic dependence.

The future industrial systems are often complex with numerous components interconnected in different ways, such as in series, parallel, and bridge connections [31–34]. The proposed models for series structures may no longer be usable. Vu et al. [12, 35, 36] proposed dynamic grouping models based on a Genetic Algorithm (GA) that can be applied to multi-component systems with complex structures. In this chapter, we aim to present the potential benefits of these models for FIS maintenance planning. We demonstrate the application of grouping models to a 16-component system and explain how they can update the maintenance plan dynamically when dynamic contexts such as system reconfiguration arise.

3 Maintenance Modeling and Assumptions

3.1 System Architecture and Failure Modeling

The future industrial systems can be considered as multi-component systems consisting of n components. The Reliability Block Diagram (RBD), a success-oriented network that shows the logical connections (AND/OR) between system components required to accomplish a specified function [37], is used to model the components' interconnections. Figure 2 shows an example of the RBD of a 16-component system.

The FIS are complex systems composed of two types of components. The first type is critical, meaning that if one of these components shuts down for any reason, the entire system will shut down, e.g., components 1, 2, 6, 7 in Fig. 2. The second type is non-critical, meaning that even if one of these components stops functioning, the system can continue to operate.

The Weibull distribution law is used to describe the failure behaviors of FIS components during the planning horizon. The failure rate of component i is calculated as

$$r_i(t) = \frac{\beta_i}{\lambda_i}\left(\frac{t}{\lambda_i}\right)^{\beta_i - 1} \tag{1}$$

where, $\lambda_i > 0$ is the scale parameter, and β_i is the shape parameter of the Weibull distribution. The estimation of these parameters can be derived from the analysis of historical failure data and through the execution of the prognostic step.

3.2 Maintenance Activities and Costs

We consider two types of maintenance activities: *(1)* corrective maintenance (CM) is performed immediately following a component failure to restore the component

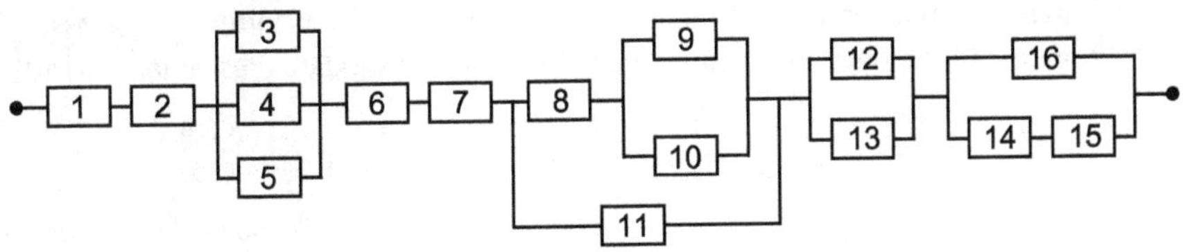

Fig. 2 RBD of a 16-component system

to its previous state (minimal repair); *(2)* preventive maintenance (PM) should be planned to replace the component by a new one after a certain period since the longer a component is in operation, the higher its failure rate will be. The duration of both corrective and preventive maintenance can be disregarded when compared to the planning horizon duration.

When conducting preventive maintenance (PM) on component i, the following costs should be taken into account: a setup cost S, which includes crew travel expenses and preparation costs such as erecting scaffolding or opening a machine [38]; a specific cost denoted as c_i^p, which depends on the behavior and characteristics of component i; the planned shutdown cost C_{sys}^p, which must be paid if component i is critical, as executing preventive maintenance on a critical component causes the system to shut down. Consequently, we have

$$C_i^p = S + c_i^p + \pi_i \cdot C_{sys}^p \tag{2}$$

where π_i stands for an indicator function. It is equal to 1 if component i is critical, and 0 if component i is non-critical.

When a corrective action is conducted for component i, it incurs a corrective maintenance cost

$$C_i^c = S + c_i^r + \pi_i \cdot C_{sys}^u \tag{3}$$

In the given equation, c_i^r represents a specific repair cost, and C_{sys}^u denotes the unplanned shutdown cost.

We should consider the possibilities of maintaining several components at once (grouping maintenance) to share preparation tasks and save on setup costs (S) during the maintenance decision-making.

3.3 System Reconfiguration

The industrial systems of the future are confronting unprecedented challenges stemming from unpredictable and dynamic market changes. These fluctuations can significantly impact production requirements, including production demands, product design specifications, and the integration of new technologies. Consequently, the system's structure and component count may undergo modifications over time (system reconfiguration) [39]. When system reconfiguration takes place, the system RBD and the critical role of the components are subject to alteration. Consequently, it is essential to dynamically update the maintenance plan to accommodate these changes [40].

4 Predictive Maintenance Strategy for Future Industrial Systems

This section presents a dynamic approach to predictive maintenance for FIS. It considers system configuration complexity, component setup cost sharing, and system reconfiguration. The approach has been developed based on the rolling horizon, originally introduced in [23], to optimize maintenance in series configuration systems. This dynamic approach encompasses four main phases: system analysis, tentative planning, optimization through grouping using genetic algorithms, and updating of maintenance planning. A flowchart illustrating the proposed approach can be found in Fig. 3.

4.1 Phase 1: System Structure Analysis

During this phase, the system undergoes analysis to ascertain the significance of individual components or groups of components concerning a specified system function. This phase encompasses three distinct steps, as illustrated in Fig. 4.

- Step 1: Construct the RBD of the system.
- Step 2: Identify all minimal cut sets by analyzing the RBD. A minimal cut set comprises the minimal number of component failures required to cause a system failure. Additionally, a minimal cut set of the order e $(1 \leq e \leq n)$ indicates that it contains exactly e components.
- Step 3: Identify the role of components/groups of components. A component is deemed critical ($\pi_i = 1$) if it constitutes a first-order minimal cut set ($e = 1$); otherwise, it is classified as non-critical ($\pi_i = 0$). Regarding a group of components

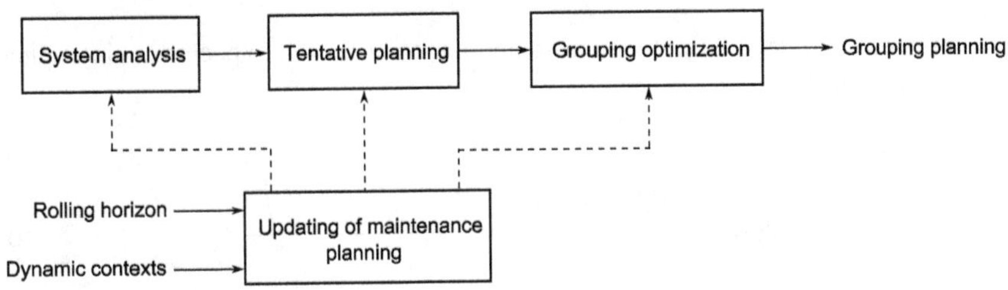

Fig. 3 Dynamic predictive maintenance approach

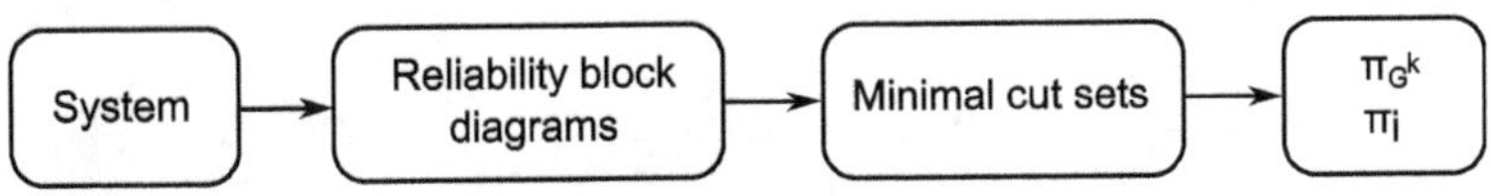

Fig. 4 System structure analysis phase

denoted as G^k, it is considered a critical group ($\pi_{G^k} = 1$) if it encompasses at least one minimal cut set. Conversely, G^k is categorized as a non-critical group ($\pi_{G^k} = 0$) if it encompasses no minimal cut set.

Assuming that the failed components are repaired immediately and instantaneously, the criticality of a component or a group of components does not change, i.e., π_i and π_G^k remain unchanged unless the system structure is altered due to system reconfiguration.

4.2 *Phase 2: Tentative Planning*

The objective of this phase is to develop an initial maintenance plan based on the optimal maintenance cycle of components. To achieve this, we first need to determine the optimal maintenance cycle.

4.2.1 Optimal PM Cycle

Let $M_i(x_i)$ represent the expected CM cost of component i in x_i time units after the last PM activity. $M_i(x_i)$ can be formulated as

$$M_i(x_i) = C_i^c \cdot \int_0^{x_i} r_i(y)dy \tag{4}$$

From Eqs. 1 and 4, we can derive

$$M_i(x_i) = C_i^c \cdot \left(\frac{x_i}{\lambda_i}\right)^{\beta_i} \tag{5}$$

If component i undergoes preventive maintenance at x_i, the expected maintenance cost rate within $[0, x_i]$ is

$$\phi(x_i) = \frac{C_i^p + M_i(x_i)}{x_i} = \frac{C_i^p + C_i^c \cdot \left(\frac{x_i}{\lambda_i}\right)^{\beta_i}}{x_i} \tag{6}$$

The optimal PM cycle represented as x_i^*, can be found by minimizing the expected maintenance cost rate $\phi(x_i)$.

$$\left.\frac{d\phi(x_i)}{dx_i}\right|_{x_i=x_i^*} = 0 \Leftrightarrow C_i^c \cdot (\beta_i - 1) \cdot \left(\frac{x_i}{\lambda_i}\right)^{\beta_i} - C_i^p = 0 \tag{7}$$

We have then

$$x_i^* = \lambda_i \cdot \sqrt[\beta_i]{\frac{C_i^p}{C_i^c \cdot (\beta_i - 1)}} \tag{8}$$

and

$$\phi_i^* = \frac{C_i^p \cdot \beta_i}{x_i^* \cdot (\beta_i - 1)} \tag{9}$$

The minimal long-term expected average cost per unit time of the system, given optimal and separate maintenance of all components, can be expressed as

$$\phi_{sys}^* = \sum_{i=1}^{n} \phi_i^*. \tag{10}$$

The optimal maintenance cycle x_i^* will be utilized to establish the tentative maintenance schedule for component i within the planning horizon.

4.2.2 Planning Horizon and PM Dates

To proceed to the subsequent phase, it is imperative to establish a short-term planning horizon. This necessitates setting the commencement of the planning horizon, denoted as T_{begin}, at the current date. The termination date, denoted as T_{end}, can be delineated as follows:

$$T_{end} = \max_i t_{i^1} \tag{11}$$

Where t_i^1 represents the initial scheduled maintenance time for component i within the planning horizon. t_i^1 is calculated as follows:

$$t_{i^1} = (x_i^* - d_i) + T_{begin} \tag{12}$$

where the total operational duration of component i at T_{begin} is represented by d_i.

If we define i^j as the jth occurrence of a PM activity for component i within the planning horizon, then we can calculate the execution time of the PM activity i^j, denoted as t_{i^j}, using the following equation:

$$t_{i^j} = t_{i^{j-1}}^* + x_i^*, \quad j > 1, \ t_{i^j} \le T_{end} \tag{13}$$

where $t_{i^{j-1}}$ denotes the tentative execution time of activity i^{j-1}.

In Fig. 3, an example of tentative planning for a three-component system is illustrated. The planning horizon, defined with $T_{begin} = 0$ and $T_{end} = \max_i t_{i^1} = t_{2^1}$, involves preventive maintenance for components 1 and 2 at times t_{1^1} and t_{2^1}, and three preventive maintenance activities for component 3 at times t_{3^1}, t_{3^2}, and t_{3^3}. Consequently, there are a total of 5 preventive maintenance activities (Fig. 5).

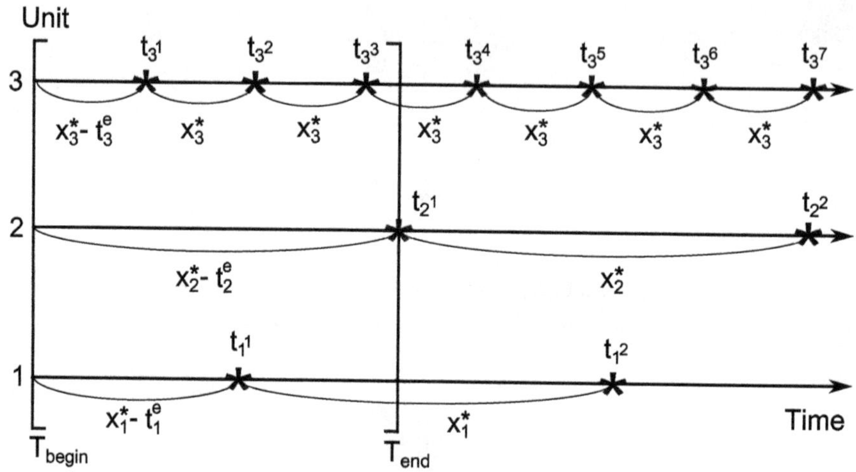

Fig. 5 An example of the tentative planning of a three components system

4.3 Phase 3: Maintenance Grouping Optimization

The primary objective of this phase is to endeavor to execute several PM activities concurrently, with the aim of minimizing the total maintenance costs.

4.3.1 Economic Profit of One Group

When all PM activities in group G^k are executed simultaneously at a specific time t_G, it results in a reduction of setup costs.

$$U_{G^k} = [card(G^k) - 1] \cdot S \tag{14}$$

In the above equation, card(G^k) represents the number of PM activities in G^k.

The process of grouping also entails a change in the maintenance date from the initial (t_{ij}) to the grouped one (t_{G^k}). As a result of this alteration, the following changes may occur [40]:

- A changing in expected deterioration costs: $M_i(t_{G^k}) - M_i(x_i^*)$;
- A changing cost arises from the delay of all future execution times after t_{ij}, calculated as $(t_{G^k} - t_{ij}) \cdot \phi_i^*$.

Let $\triangle H_{G^k}^1(t)$ represent the total penalty costs resulting from changes in the maintenance date of all PM activities in group G^k. The value of $\triangle H_{G^k}^1(t)$ can be computed using the following formula:

$$\triangle H^1_{G^k}(t) = \sum_{i^j \in G^k} [M_i(t_{G^k}) - M_i(x_i^*) - (t_{G^k} - t_{i^j}) \cdot \phi_i^*] \tag{15}$$

The optimal execution time of group G^k can be found when $\triangle H^1_{G^k}$ reaches its minimal value $\triangle H^{1*}_{G^k}$, denoted as $t^*_{G^k}$.

$$t^*_{G^k} = \text{argmin} t_{G^k} \triangle H^1_{G^k}(t_{G^k}) \text{ and } \triangle H^{1*}_{G^k} = \triangle H^1_{G^k}(t^*_{G^k}) \tag{16}$$

Finally, if the group G^k is critical, the grouping can lead to an additional gain or penalty cost associated with system shutdown.

$$\triangle H^2_{G^k} = \left(\sum_{i^j \in G^k} \pi_i - \pi_{G^k} \right) \cdot C^p_{sys} \tag{17}$$

From (14), (16), and (17), the economic profit of group G^k can be calculated as

$$EP(G^k) = U_{G^k} - \triangle H^{1*}_{G^k} + \triangle H^2_{G^k}. \tag{18}$$

4.3.2 Total Economic Profit

In order to minimize maintenance costs, we need to consider all potential groups within the planning horizon. To achieve this, we define a grouping solution (GS) as a set of m mutually exclusive groups, denoted as G^1 to G^m, which collectively encompass all N PM activities within the planning horizon.

$$G^l \cap G^k = \emptyset, \ \forall l \neq k \text{ and } G^1 \cup G^2 \cup ... \cup G^m = \{1, ..., N\}. \tag{19}$$

The total economic profit of a grouping solution (GS), referred to as EPS, is determined using the following formula:

$$EPS(GS) = \sum_{G^k \in GS} EP(G^k) = \sum_{G^k \in GS} (U_{G^k} - \triangle H^{1*}_{G^k} + \triangle H^2_{G^k}). \tag{20}$$

A grouping solution or plan is optimal when total economic profit reaches its maximum value, i.e.,

$$GS^* = \arg\max_{GS} EPS(GS) \tag{21}$$

The problem of determining the optimal grouping plan cannot be resolved using analytical methods. In Sect. 4, we will introduce a solution based on Genetic algorithms (GA).

4.4 Phase 4: Updating the Maintenance Planning

4.4.1 Rolling Horizon

The maintenance plan formulated in the previous phase can be utilized as an optimal strategy for the timeframe between T_{begin} and T_{end}. However, upon the expiration of this planning horizon, a new maintenance plan will be necessitated for the subsequent horizon. To achieve this, the rolling horizon procedure is employed. This process entails reverting to phase 2 of the optimization scheme, redefining all preventive maintenance activities for the new prospective planning horizon, and subsequently iterating the process.

4.4.2 Update the Maintenance Planning Due to System Reconfiguration

The system structure remains unchanged, so there's no need to re-analyze it. However, changes are made at the component level, requiring us to revisit phase 2 to update these changes and so on.

5 Maintenance Optimization Using Genetic Algorithms

5.1 Complexity of the Grouping Optimization Problem

The Stirling numbers of the second kind (Rennie 1969), denoted as $S(N, k)$, represent the count of ways to divide a set of N objects into k non-empty subsets. In simple terms, $S(N, k)$ provides the total number of grouping solutions that consist of exactly k groups. This number can be calculated using the following equation:

$$S(N, k) = \frac{1}{k!} \cdot \sum_{j=0}^{k-1} (-1)^j \cdot \binom{k}{j} \cdot (k - j)^N \tag{22}$$

It's important to remember that the maximum number of groups in a solution where you group N PM activities is $N - 1$. If the number of groups equals the number of PM activities, then no grouping is performed. The total possible grouping solutions, denoted as $S(N)$, can be calculated as:

$$S(N) = \sum_{k=1}^{N-1} S(N, k) \tag{23}$$

The problem of finding the best grouping solution from $S(N)$ possible solutions is known to be NP-complete [41]. Using an exact method that considers all possible grouping solutions takes too much computational time, particularly for high values of N. To address this issue, we will introduce a solution based on genetic algorithms (GA) in the next subsection [42].

5.2 *Implementation of GA*

The Genetic Algorithm (GA) begins by generating an initial population of potential solutions. Each solution is evaluated using an objective function. Genetic operations, such as crossover and mutation, are then applied to explore potentially improved solutions at each iteration. The algorithm continues until the maximum number of iterations is reached. In the next paragraph, we will provide more details about the main steps of the GA and some modifications made when it is applied to the grouping optimization problem.

Coding. A coding method defines how a solution is encoded in the algorithm. In this particular problem, each solution is depicted by an array X containing N elements that correspond to N PM activities. The array X is created using the following coding function:

$$i^j \in G^k \Rightarrow X(i^j) = k \ ; \ k \in [1, .., (N-1)] \tag{24}$$

Generation of initial population. We start by creating an initial set of solutions (initial population) containing n_{pop} solutions. To create a grouping solution, we randomly choose the number of groups between 1 and N-1. Then, we randomly assign all maintenance activities to the chosen groups.

Calculation of fitness values. The fitness value of a solution X is

$$f(X) = EPS(GS_X) \tag{25}$$

where GS_X represents the grouping solution for the solution X, and $EPS(GS_X)$ denotes the total economic profit of GS_X. The value of $EPS(GS_X)$ is determined using Eq. 20.

Crossover. The "linear ranking selection" method [43] is utilized to select parent solutions for crossover. Once the solutions are selected, a single-point crossover process is applied to generate improved solutions for the next generation. In this process, a random location is chosen, and subsequences before and after that location are exchanged between the two selected solutions to produce two children. An example demonstrating how the single-point crossover works is shown in Fig. 6a.

Mutation. To perform the mutation, each selected solution involves moving a random PM activity from one group to another to generate a new solution. An example of a mutation operator is shown in Fig. 6b.

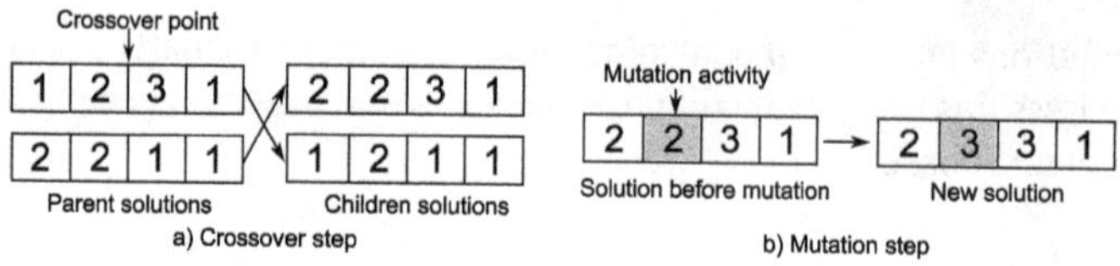

Fig. 6 Illustration of crossover and mutation steps

6 Numerical Example

We are using a 16-component FIS with the RBD shown in Fig. 2. The failure behavior of its components during the planning horizon is estimated and modeled by Weibull distributions with scale parameters $\lambda_i > 0$, and shape parameters $\beta_i > 1$ (Table 1). In Table 1, the specific PM and CM costs (c_i^p, c_i^r), as well as the operational time of the components at the beginning of the planning horizon (d_i), are reported. We have set $S = 30$, $C_{sys}^u = 70$, and $C_{sys}^p = 60$ for the other costs.

In the next subsections, we will illustrate how predictive maintenance can be used in this system and how to update the maintenance plan after a system reconfiguration.

6.1 *Predictive Maintenance Planning for the 16-Component FIS*

Individual maintenance planning. We start by examining the Reliability Block Diagram (RBD) of the system and determining the criticality of each component (π_i). To establish a preliminary plan, we calculate C_i^p, C_i^c, x_i^*, and ϕ_i^* by plugging in the provided data and π_i into Equations 1 to 4 respectively. The outcomes of these calculations are summarized in Table 2.

Based on the values of d_i and x_i^*, we determine the start time (t_{i^1}) of the initial preventive maintenance (PM) activity for each component using Eq. 12. The planning horizon ranges from $T_{begin} = 0$ to $T_{end} = \max t_{i^1} = t_{15^1} = 335$. We calculate the dates for the recurring PM activities using Eq. 13, but all of the second execution

Table 1 Data of FIS' components

i	1	2	3	4	5	6	7	8	9	10	11	12	13	14	15	16
λ_i	170	250	190	200	108	285	209	97	84	149	190	117	205	241	300	235
β_i	1.55	1.75	2.00	2.00	1.70	2.00	1.95	1.75	1.50	1.50	2.00	1.70	1.75	1.75	1.80	2.00
c_i^p	205	105	245	235	385	345	225	345	345	165	345	305	225	105	105	225
c_i^r	120	72	98	107	152	80	102	50	76	92	78	66	96	52	92	65
d_i	220	246	116	108	31	306	105	26	44	123	169	61	96	88	24	85

Table 2 Results of the individual maintenance optimization

i	1	2	3	4	5	6	7	8	9	10	11	12	13	14	15	16
π_i	1	1	0	0	0	1	1	0	0	0	0	0	0	0	0	0
C_i^p	295	195	275	265	415	435	315	375	375	195	375	335	255	135	135	255
C_i^c	220	172	128	137	182	180	202	80	106	122	108	96	126	82	122	95
x_i^*	302	316	278	278	216	443	269	276	309	323	354	301	361	378	359	385
ϕ_i^*	85	70	163	170	185	137	165	250	266	200	185	240	266	290	335	300

times (t_{i^2}) fall outside the planning horizon. As a result, each component receives preventive maintenance only once within the planning horizon (with $N = n = 16$). The average maintenance cost rate of the system, assuming all PM activities are executed separately, is:

$$\phi_{sys} = \sum_{i=1}^{n} \phi_i^* = 35.17 \tag{26}$$

The total maintenance cost within the planning horizon is then

$$TC = \phi_{sys} \cdot (T_{end} - T_{begin}) = 11782 \tag{27}$$

Grouping maintenance planning. The genetic algorithm was used to identify the most effective grouping solution for creating an optimal maintenance plan and reducing maintenance costs.

The optimal maintenance grouping plan consists of three groups. The composition of each group, their execution dates, and their criticality are shown in Table 3. This grouping solution has resulted in an economic profit of $EPS = 523$. When compared to individual maintenance planning, the grouped maintenance leads to cost savings of

$$\frac{EPS}{TC} \cdot 100\% = 4.45\% \tag{28}$$

Finally, the maintenance plans for both individuals and groups are represented in Fig. 7.

Table 3 Optimal solution

Groups	1	2	3
PM activities	$\{1^1, \ldots, 7^1, 10^1, 11^1\}$	$\{8^1, 9^1, 12^1, 16^1\}$	$\{13^1, 14^1, 15^1\}$
t_{G^k}	160	258	290
$EP(G^k)$	381	85	57
π_{G^k}	1	0	0

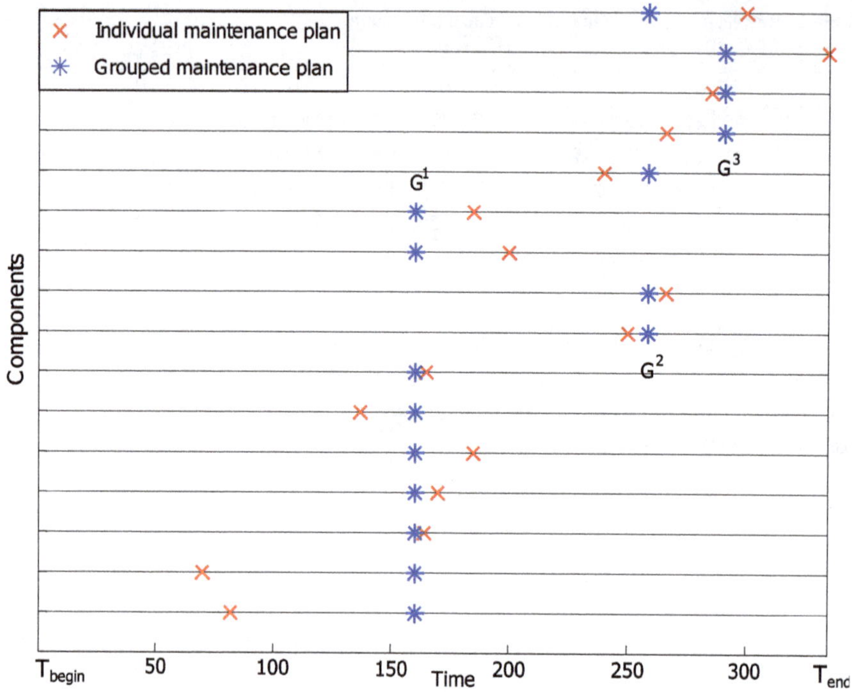

Fig. 7 Individual and grouping plans

6.2 *Updating of the Maintenance Plans Due to System Reconfiguration*

Here, we outline the process of updating maintenance planning while taking into account a system reconfiguration action. We assume that at time $t_{reconf} = 200$, components 17 and 18 have been added to the system in response to increasing customer demands (refer to Fig. 8).

The components' parameters are given in Table 4.

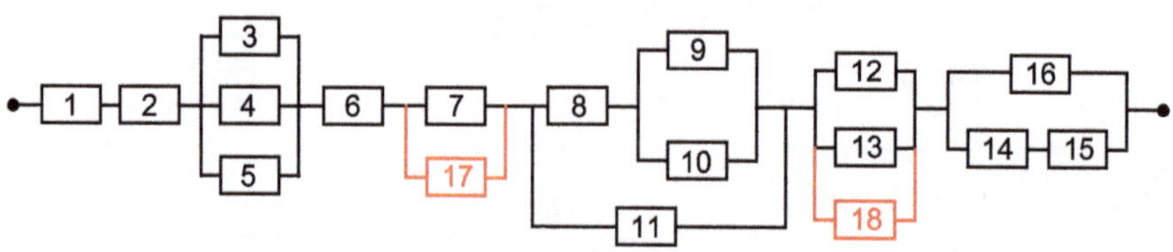

Fig. 8 RBD of the FIS after reconfiguration

Table 4 Parameters of components 17 and 18

Components	λ_i	β_i	c_i^p	c_i^r	d_i
17	150	1.95	400	120	0
18	110	2.30	250	100	0

Table 5 Update of the optimal PM cycles

Components	7	17	18
C_i^p	255	430	280
C_i^c	132	150	130
x_i^*	301	264	137
ϕ_i^*	1.74	3.34	3.62

Due to system reconfiguration, groups G^2 and G^3 scheduled for execution after t_{reconf} may no longer be optimal. We need to revisit the system analysis phase to update the maintenance plan.

Individual maintenance planning considering the system reconfiguration. To update individual maintenance planning, we need to reassess the system to update the criticality of its components. The obtained results show that component 7 becomes a non-critical one after the reconfiguration.

The individual optimization is performed to calculate the optimal preventive maintenance (PM) cycles for components 17 (x_{17}^*) and 18 (x_{18}^*). The optimal PM cycle for component 7 (x_7^*) is also updated due to changes in its criticality. The results are presented in Table 5.

As per convention, when $t_{reconf} = 200$, it serves as the commencement of a new planning horizon ($T'_{begin} = 200$). The operational time of component i at the new starting date T'_{begin} is then denoted by d'_i and updated as follows:

- with components of group G^1, which were maintained at $t_{G^1} = 160$, the operational time of these components is calculated by the following equation:

$$d'_i = (T'_{begin} - T_{begin}) - t_{G^1},\ i \in G^1;$$

- with components of groups G^2 and G^3:

$$d'_i = (T'_{begin} - T_{begin}) + d_i,\ i \in G^2\ or\ i \in G^3;$$

- with new components 17 and 18: $d_{17} = d_{18} = 0$.

The first tentative maintenance time for all components was calculated based on operational time and reported in Table 6. A new planning horizon is set between $T'_{begin} = 200$ and $T'_{end} = t_{6^1} = 603$. The dates for recurring preventive maintenance

Table 6 Updated tentative maintenance dates

Components	1	2	3	4	5	6	7	8	9
$d_i^{'}$	40	40	40	40	40	40	40	226	244
x_i^*	302	316	278	278	216	443	300	276	309
t_{i1}	462	477	438	438	376	603	461	250	266.00
t_{i2}					593			526	575
Components	10	11	12	13	14	15	16	17	18
$d_i^{'}$	40	40	261	295	288	224	285	0	0
x_i^*	323	354	301	362	378	359	385	264	137
t_{i1}	484	514	240	266	290	335	300	464	337
t_{i2}			541						574

Table 7 Optimal grouping solution after the system reconfiguration

Groups	PM activities	t_{G^k}	$EP(G^k)$	π_{G^k}	EPS
1	$\{8^1, 9^1, 12^1, 14^1, 16^1\}$	260	115	0	599
2	$\{5^1, 13^1, 15^1, 18^1\}$	344	71	0	
3	$\{1^1, ..., 4^1, 6^1, 7^1, 10^1, 11^1, 17^1, 18^2\}$	475	338	1	
4	$\{5^2, 8^2, 9^2, 12^2\}$	565	76	0	

(PM) activities are then calculated. It's important to note that within this new planning horizon, components 5, 8, 9, 12, and 16 will undergo preventive maintenance twice. In this case, the total number of PM activities is greater than the number of components ($N = 23 > n = 18$).

For the new system, the average maintenance cost rate when all PMs are performed separately is

$$\phi_{sys} = \sum_{i=1}^{18} \phi_i^* = 41.47 \tag{29}$$

The total maintenance cost in the new planning horizon is

$$TC = \phi_{sys} \cdot (T_{end}^{'} - T_{begin}^{'}) = 16722 \tag{30}$$

Grouped maintenance planning. We apply GA again to find the optimal grouping planning. The obtained results are reported in Table 7.

The total economic profit is equal to $EPS = 599$. When compared to the individual maintenance planning, the grouping maintenance leads to a saving cost of

$$\frac{EPS}{TC} \cdot 100\% = 3.59\% \tag{31}$$

The tentative and grouping maintenance plans after the system reconfiguration are shown in Fig. 9.

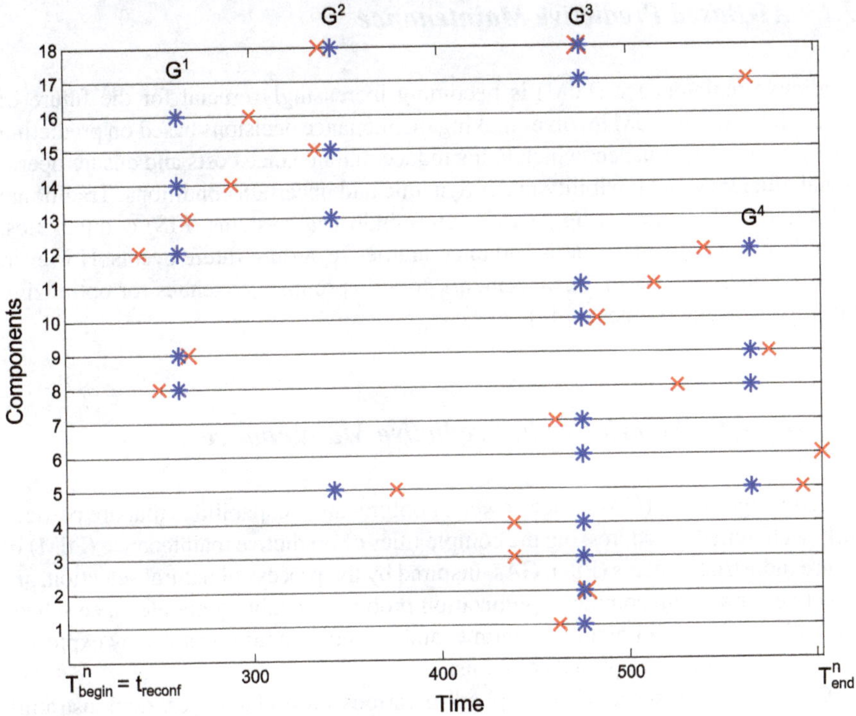

Fig. 9 Tentative and grouping maintenance plans after the system reconfiguration

7 Perspectives for Artificial Intelligence-Based Predictive Maintenance Optimization

Integrating Artificial Intelligence (AI) in predictive maintenance (PdM) is revolutionizing the landscape of industrial maintenance. AI-driven PdM leverages advanced data analytics to predict equipment failures, optimize maintenance schedules, reduce downtime, and extend the lifespan of assets. This approach is critical for future industrial systems (FIS), which are characterized by their complexity and dynamic nature. Traditional PdM methods often fail to address the uncertainties and interactions inherent in FIS, whereas AI technologies offer robust solutions through advanced modeling and predictive capabilities.

One of the most promising perspectives is the use of Explainable AI (XAI). XAI enhances the transparency of AI models, making them more understandable to human operators. This is crucial in maintenance contexts where trust and clarity in AI-driven decisions can significantly improve acceptance and implementation. Another key perspective is the application of Reinforcement Learning (RL). RL algorithms continuously learn from operational data and feedback, adapting real-time maintenance strategies to evolving conditions.

7.1 AI-Based Predictive Maintenance

Predictive maintenance (PdM) is becoming increasingly crucial for the future of industrial systems. PdM involves making maintenance decisions based on predicting the system's performance, which helps reduce maintenance costs and ensure operational efficiency and flexibility under dynamic and uncertain conditions. Traditional PdM approaches often struggle with future industrial systems (FIS) complexities, including complex interactions and uncertainties regarding future events. However, artificial intelligence (AI) advancements provide promising avenues for optimizing PdM in these environments [44].

7.2 Genetic Algorithms in Predictive Maintenance

Genetic Algorithms (GAs) offer powerful optimization capabilities that are particularly well-suited for addressing the complexities of predictive maintenance (PdM) in future industrial systems (FIS). GAs, inspired by the process of natural selection, are effective for solving complex optimization problems, enabling maintenance schedule optimization, minimizing downtime, and extending equipment life by exploring vast search spaces for optimal solutions.

GAs have been successfully applied to various PdM challenges, demonstrating significant improvements in maintenance strategies and system reliability.

- **Maintenance Scheduling**: GAs optimize the timing and frequency of maintenance activities to minimize costs and downtime while ensuring high equipment reliability [45].
- **Resource Allocation**: GAs optimize the allocation of maintenance resources such as labor and spare parts, ensuring efficient use of resources and reducing overall maintenance budgets [46].

7.3 Future Research Directions

- **Computational Complexity**: GAs can be computationally intensive, especially for large-scale problems with many variables and constraints. Future research should focus on developing more efficient GA algorithms and parallel processing techniques to address this challenge.
- **Data Quality**: The performance of GAs is highly dependent on the quality of input data. Accurate and high-resolution data are crucial for generating reliable maintenance schedules [47].

- **Integration with Other AI Techniques**: Combining GAs with other AI techniques, such as machine learning and reinforcement learning, can enhance their optimization capabilities. Hybrid approaches can leverage the strengths of each technique, providing more robust and adaptive PdM solutions [48].

8 Conclusions

In this chapter, we introduce a predictive maintenance approach for future industrial systems that considers the sharing of setup costs, the complexity of system structure, and the system reconfiguration. The optimal predictive grouping plan was established with the help of a genetic algorithm. A numerical example was proposed to illustrate the functioning of the approach during normal and reconfiguration situations. The obtained results provide strong evidence of the exceptional dynamism and effectiveness of the predictive maintenance approach, making it a promising solution for the maintenance planning of future industrial systems. It's important to note that, in addition to maintenance costs, other objectives such as reliability, maintainability, safety, and sustainability are also crucial for future industrial systems. Our upcoming research will focus on developing a predictive approach that considers all of these objectives during maintenance planning. As the problem becomes more complex, we may need to use multi-objective optimization approaches and other AI techniques such as reinforcement learning.

References

1. Hermann M, Pentek T, Otto B (2016) Design principles for industrie 4.0 scenarios. In: 2016 49th Hawaii international conference on system sciences (HICSS), pp 3928–3937. IEEE
2. Kim NH, An D, Choi JH (2016) An Introduction. Prognostics and health management of engineering systems. Springer
3. Wang H (2002) A survey of maintenance policies of deteriorating systems. Eur J Oper Res 139(3):469–489
4. Pinciroli L, Baraldi P, Zio E (2023) Maintenance optimization in industry 4.0. Reliab Eng & Syst Saf 234:109204
5. De Jonge B, Scarf PA (2020) A review on maintenance optimization. Eur J Oper Res 285(3):805–824
6. Ding S-H, Kamaruddin S (2015) Maintenance policy optimization-literature review and directions. Int J Adv Manuf Technol 76:1263–1283
7. Keizer MCAO, Flapper SDP, Teunter RH (2017) Condition-based maintenance policies for systems with multiple dependent components: a review. Eur J Oper Res
8. Wang Y, Li X, Chen J, Liu Y (2022) A condition-based maintenance policy for multi-component systems subject to stochastic and economic dependencies. Reliab Eng & Syst Saf 219:108174
9. Vu HC, Do P, Fouladirad M, Grall A (2020) Dynamic opportunistic maintenance planning for multi-component redundant systems with various types of opportunities. Reliab Eng & Syst Saf 198:106854

10. Wocker M, Betz NK, Feuersänger C, Lindworsky A, Deuse J (2020) Unsupervised learning for opportunistic maintenance optimization in flexible manufacturing systems. Procedia CIRP 93:1025–1030
11. Shi H, Zhang J, Zio E, Zhao X (2023) Opportunistic maintenance policies for multi-machine production systems with quality and availability improvement. Reliab Eng & Syst Saf 234:109183
12. Vu HC, Do P, Barros A (2016) A stationary grouping maintenance strategy using mean residual life and the birnbaum importance measure for complex structures. IEEE Trans Reliab 65(1):217–234
13. Wang J, Zhang X, Zeng J (2022) Dynamic group-maintenance strategy for wind farms based on imperfect maintenance model. Ocean Eng 259:111311
14. Shi D, Ma H, Ma C (2023) A dynamic maintenance strategy for multi-component systems using a genetic algorithm. CMES-Comput Model Eng & Sci 134(3)
15. Bris R, Châtelet E, Yalaoui F (2003) New method to minimize the preventive maintenance cost of series-parallel systems. Reliab Eng & Syst Saf 82(3):247–255
16. Samrout M, Yalaoui F, Chatelet E, Chebbo N (2005) New methods to minimize the preventive maintenance cost of series-parallel systems using ant colony optimization. Reliab Eng & Syst Saf 89(3):346–354
17. Pereira CMNA, Lapa CMF, Mol ACA, Da Luz AF (2010) A particle swarm optimization (PSO) approach for non-periodic preventive maintenance scheduling programming. Progr Nuclear Energy 52(8):710–714
18. Gertsbakh IB (1984) Optimal group preventive maintenance of a system with observable state parameter. Adv Appl Probab 16(4):923–925
19. Vergin RC, Scriabin M (1977) Maintenance scheduling for multicomponent equipment. AIIE Trans 9(3):297–305
20. Love CE, Guo R (1996) Utilizing weibull failure rates in repair limit analysis for equipment replacement/preventive maintenance decisions. J Oper Res Soc 47(11):1366–1376
21. Sheu SH, Jhang JP (1997) A generalized group maintenance policy. Eur J Oper Res 96(2):232–247
22. Park M, Pham H (2012) A generalized block replacement policy for a k-out-of-n system with respect to threshold number of failed components and risk costs. IEEE Trans Syst Man Cybern Part A: Syst Hum 42(2):453–463
23. Wildeman RE, Dekker R, Smit ACJM (1997) A dynamic policy for grouping maintenance activities. Eur J Oper Res 99(3):530–551
24. Bouvard K, Artus S, Bïrenguer C, Cocquempot V (2011) Condition-based dynamic maintenance operations planning & grouping. application to commercial heavy vehicles. Reliab Eng & Syst Saf 96(6):601–610
25. Van PD, Barros A, Bïrenguer C, Bouvard K, Brissaud F (2013) Dynamic grouping maintenance with time limited opportunities. Reliab Eng & Syst Saf 120:51–59
26. Do Van P, Vu HC, Barros A, Berenguer C (2012) Grouping maintenance strategy with availability constraint under limited repairmen. In: 8th IFAC symposium on fault detection supervision and safety of technical processes, 29–31 August, Mexico city, Mexico, pp 486–491
27. Vu HC, Van Do P, Barros A, Berenguer C (2021) Maintenance activities planning and grouping for complex structure systems. In: Avance in safety, reliability and risk management - Proceedings of ESREL, 25–29 June 2012. Helsinki, Finland, p 2012
28. Vu HC, Do Van P, Barros A, Berenguer C (2012) Dynamic grouping maintenance for complex structure systems with non-negligible replacement time. In: Workshop on advanced maintenance engineering, services and technology - Proceedings of A-MEST 2012, 22–23 November 2012, Seville, Spain, pp 79–84
29. Van Horenbeek A, Pintelon L (2013) A dynamic predictive maintenance policy for complex multi-component systems. Reliab Eng & Syst Saf. http://dx.doi.org/10.1016/j.ress.2013.02.029
30. Omshi EM, Grall A, Shemehsavar S (2020) A dynamic auto-adaptive predictive maintenance policy for degradation with unknown parameters. Eur J Oper Res 282(1):81–92

31. Sahinoglu M, Ramamoorthy CV (2005) RBD tools using compression, decompression, hybrid techniques to code, decode, and compute reliability in simple and complex embedded systems. IEEE Trans Instrum Meas 54(5):1789–1799
32. Chen SK, Ho TK, Mao BH (2007) Reliability evaluations of railway power supplies by fault-tree analysis. Electric Power Appl IET 1(2):161–172
33. Maddikunta PKR, Pham Q-V, Prabadevi B, Deepa N, Dev K, Gadekallu TR, Ruby R, Liyanage M (2022) Industry 5.0: a survey on enabling technologies and potential applications. J Ind Inf Integr 26:100257
34. Bezoui M, Qaisar SM, Almaktoom AT, Bounceur A (2024) Advancing manufacturing efficiency: multi-objective optimization in the industry 5.0 era. In: Reality and science fiction in education
35. Vu HC, Do P, Barros A, Bérenguer C (2014) Maintenance grouping strategy for multi-component systems with dynamic contexts. Reliab Eng Syst Saf 132:233–249
36. Vu HC, Do P, Barros A (2018) A study on the impacts of maintenance duration on dynamic grouping modeling and optimization of multicomponent systems. IEEE Trans Reliab 67(3):1377–1392
37. Høyland A, Rausand M (2009) System reliability theory: models and statistical methods, vol 420. Wiley
38. Kobbacy KAH, Murthy DNP, Nicolai RP, Dekker R (2008) Optimal maintenance of multi-component systems: a review. Springer
39. Mehrabi MG, Ulsoy AG, Koren Y (2000) Reconfigurable manufacturing systems: key to future manufacturing. J Intell Manuf 11:403–419
40. Vu HC, Do P, Barros A, Bérenguer C (2014) Maintenance grouping strategy for multi-component systems with dynamic contexts. Reliab Eng Syst Saf 132:233–249
41. Garey MR, Johnson DS (1979) Computers and intractability, vol 174. Freeman, San Francisco
42. Holland JH (1984) Genetic algorithms and adaptation. Adaptive control of ill-defined systems, pp 317–333
43. Yadav SL, Sohal A (2017) Comparative study of different selection techniques in genetic algorithm. Int J Eng Sci Math 6(3):174–180
44. Hu C, Fan H, Wang Z (2022) Residual life prediction and optimal maintenance decision for a piece of equipment. Springer
45. Lapa CMF, Pereira CMNA, Mol ACDA (2000) Maximization of a nuclear system availability through maintenance scheduling optimization using a genetic algorithm. Nuclear Eng Des 196(2):219–231
46. Rankothge W, Le F, Russo A, Lobo J (2017) Optimizing resource allocation for virtualized network functions in a cloud center using genetic algorithms. IEEE Trans Netw Serv Manag 14(2):343–356
47. Sinha U, Cashman M, Cohen MB (2020) Using a genetic algorithm to optimize configurations in a data-driven application. In: Search-based software engineering: 12th international symposium, SSBSE 2020, Bari, Italy, October 7–8, 2020, Proceedings 12, pp 137–152. Springer
48. Kamel H (2011) Neural network application for structure design optimization of thin-wall structures. In: ASME international mechanical engineering congress and exposition 54952, pp 41–45

Anomaly Detection for Catalyzing Operational Excellence in Complex Manufacturing Processes: A Survey and Perspective

Moussab Orabi, Kim Phuc Tran, Sébastien Thomassey, and Philip Egger

Abstract The evolution of manufacturing processes, fueled by the transition towards Industry 5.0, has ushered in an era marked by the integration of sophisticated technologies such as the Industrial Internet of Things (IIoT), artificial intelligence (AI), and machine learning. This transformation has spawned smarter, more efficient production environments that, while advantageous, have led to the emergence of complex anomalies and deviations that traditional anomaly detection (AD) models struggle to effectively address. This chapter delves into the complexities of smart manufacturing, proposing advanced AD methodologies designed to cope with the challenges posed by the integration of cutting-edge technologies. These challenges include handling the vast volume, speed, variety, accuracy, and the inherently noisy, non-stationary, and high-dimensional nature of smart manufacturing data. By transcending simple data storage and analysis, the proposed approach links data directly to operational processes, allowing for a deeper understanding of end-to-end manufacturing activities. This not only speeds up the detection of anomalies but also aids in pinpointing the process parameters that cause quality deviations, bridging the gap between conventional models and the nuanced requirements of smart manufacturing.

Keywords Anomaly detection · Explainable artificial intelligence · Smart manufacturing · Industry 5.0 · System reliability · Deep learning · Adversarial learning · IIoT · Transformer · Multivariate time series data

M. Orabi (✉) · K. P. Tran · S. Thomassey
University of Lille, ENSAIT, ULR 2461 - GEMTEX - Génie et Matériaux Textiles, 59000 Lille, France
e-mail: Moussab.Orabi@gmail.com

M. Orabi · P. Egger
Rosenberger Hochfrequenztechnik GmbH & Co. KG, Hauptstraße 1, 83413 Fridolfing, Germany

M. Orabi · K. P. Tran
International Chair in DS & XAI, International Research Institute for Artificial Intelligence and Data Science, Dong A University, Danang, Vietnam

K. P. Tran (ed.), *Artificial Intelligence for Safety and Reliability Engineering*, Springer Series in Reliability Engineering,
https://doi.org/10.1007/978-3-031-71495-5_4

1 Introduction

The landscape of modern manufacturing has witnessed a profound transformation in recent years, characterized by the integration of advanced technologies that mark the era of Industry 4.0 and lay the groundwork for the emerging concept of Industry 5.0. This transformation is epitomized by the rise of smart manufacturing, a key component of Industry 4.0, which represents a significant shift in how industrial processes are designed, monitored, and optimized. Smart manufacturing leverages technologies such as the Internet of Things (IoT), cloud computing, big data analytics, and artificial intelligence (AI) to create an interconnected and intelligent production environment. While it is a critical element of Industry 4.0 [1], focusing on automation and data-driven processes, it also aligns with the values of Industry 5.0 [2], which emphasizes the collaboration between human ingenuity and advanced technological systems and enable seamless communication between machines, sensors, and human operators. It empowers organizations to make data-driven decisions, automate routine tasks, and achieve unprecedented levels of operational excellence. This evolution is not merely a technological advancement but a holistic approach that fosters collaboration between humans and machines. This blend of technology and human-centric approaches in smart manufacturing is reshaping the future of industrial operations, setting new standards in efficiency and innovation. As the industry gravitates towards this tech-centric approach, production processes have become increasingly complex, primarily driven by the sophisticated and evolving requirements of consumers. This complexity is not just a byproduct of demanding market forces but also a reflection of the intricate interplay of various manufacturing elements coming together in a smart factory setting. In this digital era, the manufacturing sector is inundated with data characterized by its significant size, variety, and velocity. This data, ranging from sensor readings to production logs, presents both challenges and opportunities. It requires sophisticated tools for efficient management and analysis but also holds the key to unlocking valuable insights for optimizing production processes, anticipating maintenance needs, and enhancing overall operational efficiency. This paradigm underscores the critical role of data-driven strategies in realizing the full potential of smart manufacturing in the context of Industry 4.0 and beyond. The seamless integration of vast data in smart manufacturing heralds a new era of industrial operations, where the intelligent analysis of this information becomes pivotal. AI plays a multifaceted role in smart manufacturing. Techniques such as predictive maintenance, machine learning algorithms to anticipate equipment failures before they occur, and process optimization [3], where AI models analyze production data to identify inefficiencies and suggest improvements, are increasingly common. Additionally, AI-driven robotics and automation are revolutionizing assembly lines, enabling more precise and efficient operations. AI's role in quality control is also crucial, as it helps in the accurate and rapid inspection of products, thereby ensuring high standards of quality and reducing waste. However, with the increased complexity of production processes comes the inevitable rise of anomalies. These anomalies, arising from equipment malfunctions, process deviations, or external factors, pose a significant threat to operations, potentially increasing costs and compromising product quality. This evolution has underscored the inadequacies of traditional quality control methods, like statistical analysis and Control Charts, which were once staples for ensuring quality and optimizing processes. These methods no longer meet the demands of the intricate, modern manufacturing environments. The dynamic, multivariate, and often non-stationary nature of data produced in smart factories presents challenges that these conventional tools struggle to address. Similarly, even traditional machine learning models, while more adept at handling complexity, struggle to fully address the challenging dynamics and complex interdependencies characteristic of smart manufacturing. Therefore, there is a clear need for advanced anomaly detection (AD) strategies designed for the vast data volumes, swift alterations, and intricate variable relationships prevalent in these settings. To address the intricate challenges presented by smart manufacturing, there has been a significant move towards the development of advanced AD methods. These methods leverage the cutting-edge capabilities of deep learning, utilizing transformers for handling sequential data, and employing innovative techniques in representation learning. Additionally, the advent of federated learning has enabled the collaborative training of models across various manufacturing sites, enhancing the adaptability and effectiveness of AD systems. While these advanced techniques offer superior precision in identifying and addressing anomalies, they also introduce the need for specialized expertise. This requirement underscores the evolving nature of the manufacturing landscape, where the integration of advanced technologies is critical for navigating the complexities of modern production environments. The drive towards these innovative solutions reflects a broader trend towards a more efficient, agile, and data-driven industrial future, where the timely and accurate detection of anomalies is of paramount importance. Developing effective anomaly detection (AD) strategies for smart manufacturing (MS) encompasses overcoming a range of challenges, including the management of complex data structures, interrelated attributes, unbalanced datasets, noise, outliers, and fluctuating data patterns. The successful implementation of AD is critical as manufacturing equipment and processes grow more sophisticated, highlighting the increasing necessity for real-time data collection and AI-driven anomaly detection and correction mechanisms. Despite significant advancements in AD methodologies, a crucial area of focus remains: bridging the gap between the diverse data requirements of modern smart manufacturing and the strategies and solutions currently proposed in the literature. This gap is marked by the challenges of handling the high complexity, volume, and dynamism of data typical in smart manufacturing environments. The real-world application of AD requires not only theoretical knowledge but also an understanding of the specific operational contexts and data characteristics of these industries. Contemporary research showcases a wide array of Anomaly Detection (AD) methods, such as machine learning, deep learning, outlier detection, and data mining. Yet, formulating a comprehensive strategy that effectively meets the complex and varied challenges of smart manufacturing continues to be a significant research and innovation endeavor. This disparity sets the stage for further research that bridges the academic-practical divide, ensuring that academic advancements in AD are effectively translated into practical, industry-relevant solutions. Such research must focus on developing AD methods that are not only theoretically sound but also robust and flexible enough to handle the unique challenges of real-world applications in smart manufacturing. The ongoing effort to align the capabilities of anomaly detection (AD) technologies with the dynamic and intricate needs of modern manufacturing environments underscores the importance of developing AD strategies that are as sophisticated and adaptable as the systems they aim to monitor and improve. This alignment is crucial for the success of AD applications in real-world scenarios, where the complexity and variability of data demand highly effective and adaptable solutions.

Given the above discussion, this Chapter aims to focus on developing a framework for advanced anomaly detection strategies tailored to the demands of smart manufacturing environments. This framework will incorporate novel AI techniques that better handle the unpredictable and complex nature of industrial data. By bridging the theoretical and practical aspects of anomaly detection, this chapter will provide a solid foundation for constructing more intelligent, efficient, and adaptable manufacturing systems. This work will emphasize the integration of multi-modal data sources and advanced algorithms to enhance the precision and responsiveness of AD systems. Through the exploration of cutting-edge methodologies like unsupervised machine learning, semi-supervised models, and real-time analytics, this framework aims to optimize the identification and correction of anomalies in manufacturing processes.

Further, the chapter will delve into the specifics of deploying these advanced AD methods in a real-world manufacturing setting. It will explore the challenges associated with implementing such systems, including the need for robust data governance, the importance of continuous model training and refinement, and the integration of human expertise to ensure the effectiveness of AD solutions. Practical case studies will be examined to illustrate the successful application of these methods in various manufacturing scenarios, highlighting the improvements in productivity, reduction in operational costs, and enhancements in product quality achieved through these advanced techniques.

The rest of this Chapter is structured as follows: We begin with "An Introduction to Data in Smart Manufacturing" in Sect. 2. Here, we discuss the pivotal role of multivariate time series (MVTS) data in Smart Manufacturing (SM), highlighting the intricacies and challenges associated with capturing, processing, and analyzing data from various sensors and systems in real-time. We explore how data-driven insights enhance operational efficiency and predictive maintenance strategies. Subsequently, the chapter delves into "Segmentation and Windowing in Time Series Analysis" in Sect. 3, examining different techniques to break down complex data streams into manageable and analyzable segments. This is followed by a detailed look at "Segmentation for Time Series Reconstruction and Representation", where we discuss methods for reconstructing and representing time series data to enhance anomaly detection capabilities. In "Segmentation at Characteristic Points of the Time Series", we focus on identifying key points in the data stream that significantly impact the detection of anomalies. "Locality in Time Series Analysis" in Sect. 3.3 addresses the spatial and temporal locality aspects of data analysis in smart manufacturing. The chapter continues with "Anomalies within MVTS in Smart Manufacturing" in Sect. 4, where we discuss specific types of anomalies typical to smart manufacturing environments and their implications. A step-by-step guideline for anomaly detection in time series data is provided in Sect. 5. Section 6 provides perspectives on Anomaly Detection in Smart Manufacturing and outlines some future research directions. The chapter concludes with "Conclusion" in Sect. 7, summarizing key insights and future directions in anomaly detection for smart manufacturing.

2 Time Series Data (TS) in Smart Manufacturing

Time series data, a sequence of data points collected over time intervals, plays a pivotal role in smart manufacturing. It can either be a measurement of a single process parameter, known as a univariate time series, or multiple measurements of various parameters taken simultaneously, referred to as a multivariate time series like temperature, pressure, vibration, and production rates, each recorded at specific time stamps. The univariate data can be represented as:

$$T = \{t_1, t_2, t_3, \ldots, t_n\} \tag{1}$$

where t_i denotes the measurement at time i.

The analysis of these time series data is crucial for capturing the dynamic behavior of manufacturing processes [4, 5]. To facilitate this, time series data can be decomposed into several components, each highlighting different aspects of the data. These components include:

Trend Component (T): Represents the long-term progression of the series, indicating a persistent, directional movement in the data. In a manufacturing context, this could manifest as a gradual increase in machine temperature over several months due to wear and tear. Mathematically, the trend component can be expressed as a function $T(t)$ where t is the period.

Seasonal Component (S): This captures regular fluctuations in the data that occur at specific intervals, such as daily, weekly, or monthly. For instance, a pattern in energy usage in a factory that peaks at certain times of the day and drops during others. The seasonal component can be modeled as $S(t)$, which repeats itself at regular intervals.

Cyclical Component (C): These are fluctuations occurring at irregular intervals, often influenced by broader economic or industrial factors. Unlike the seasonal component, cyclical components do not have a fixed period. This can be represented as $C(t)$, indicating cyclical effects over time.

Irregular Component (I): This includes random or unpredictable variations in the series. These could be sporadic events affecting the manufacturing process, like unexpected machine downtime or temporary supply chain disruptions. The irregular component is denoted as $I(t)$, encompassing random variations that are not explained by the trend, seasonal, or cyclical components.

In the context of smart manufacturing, each of these components plays a crucial role in understanding and forecasting the behavior of the manufacturing system [6]. For example, by analyzing the trend component, manufacturers can predict when equipment might fail or when to schedule maintenance [7]. The seasonal component can help in optimizing resource allocation during peak and off-peak production times [8].

A general model to represent these components in a time series is an **Additive Model** $Y(t)$ can be written as a combination of these elements, and used when seasonal variations are relatively constant:

$$Y(t) = T(t) + S(t) + C(t) + I(t) \tag{2}$$

where $Y(t)$ is the observed data at time t (Fig. 1). On the other hand, for some analyses, a **Multiplicative Model** as in Fig. 2, might be more appropriate, especially when the seasonal or cyclical variations increase with the level of the trend:

$$Y(t) = T(t) \times S(t) \times C(t) \times I(t) \tag{3}$$

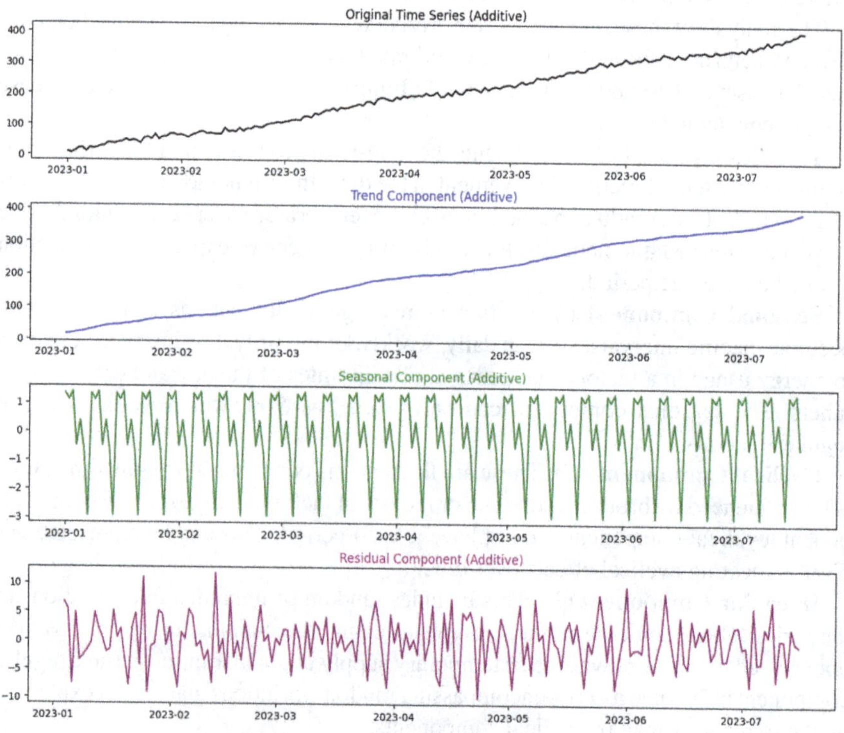

Fig. 1 Decomposition of the time series in additive components

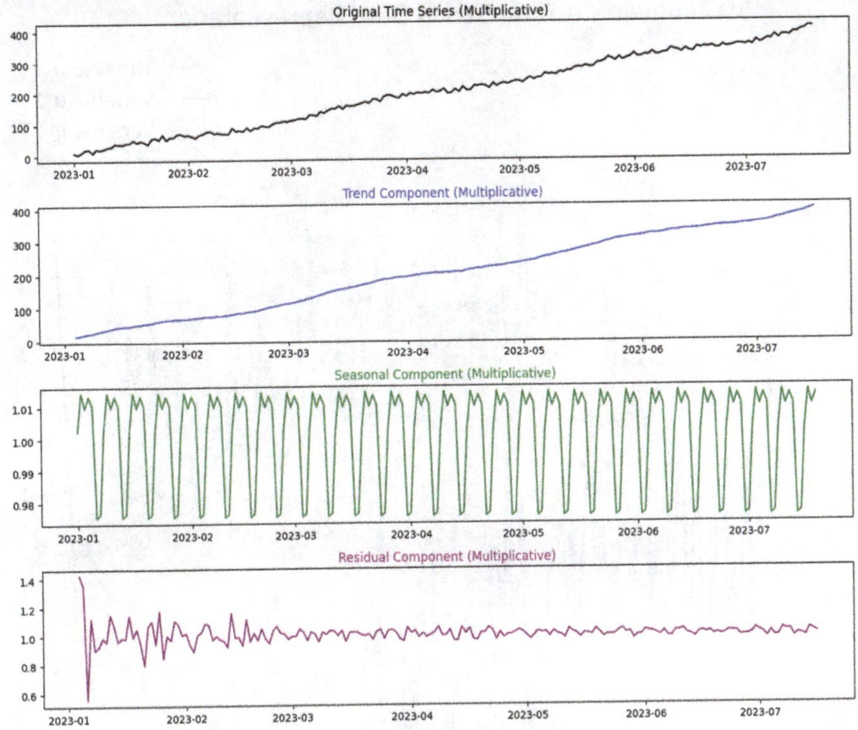

Fig. 2 Decomposition of the time series in multiplicative components

After delving into the nuances of *univariate time series* data and its components, we are now ready to venture into the realm of multivariate time series data (MTVS). *Multivariate Time Series (MVTS)* introduces a higher level of complexity by analyzing multiple time-dependent variables simultaneously. This method is especially pertinent in advanced applications like smart manufacturing, where processes are intricate and interconnected. By considering variables such as temperature, pressure, vibration, and speed together, MVTS offers a comprehensive view of production dynamics. It enables the detection of subtle patterns and anomalies that univariate analyses may overlook, providing critical insights for optimizing operations, monitoring, and quality control [9]. In such environments, the analysis of MVTS is indispensable for informed decision-making and process improvement. Mathematically, MVTS can be conceptualized as a matrix:

$$M = \begin{bmatrix} u_{11} & u_{12} & \cdots & u_{1p} \\ u_{21} & u_{22} & \cdots & u_{2p} \\ \vdots & \vdots & \ddots & \vdots \\ u_{n1} & u_{n2} & \cdots & u_{np} \end{bmatrix} \tag{4}$$

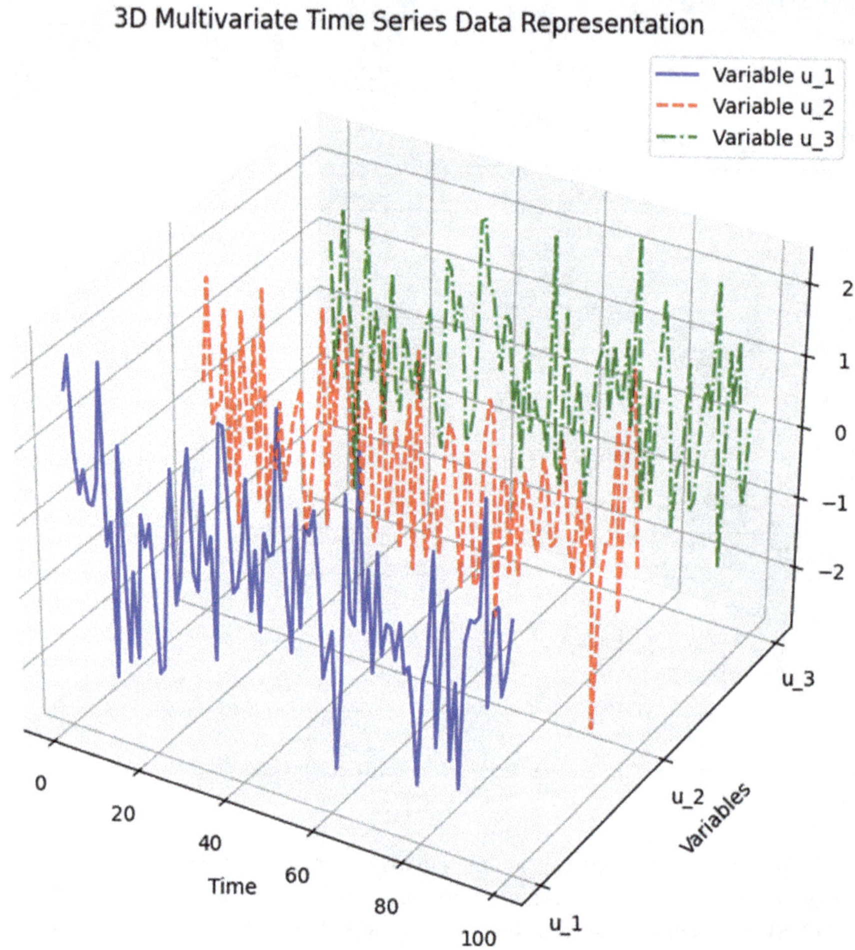

Fig. 3 Multivariate time series data representation

where u_{ij} represents the jth variable at time i, and p is the number of variables. Alternatively, it can also be expressed as:

$$X = \{x_1, x_2, ..., x_t\} \tag{5}$$

with each $x_t = [x_{t1}, x_{t2}, ..., x_{tn}]^T$ representing the values of all variables at time t, where n is the number of variables and T the total number of time points, as illustrated in Fig. 3. For example, in a production line, correlations between variables like temperature, pressure, and speed can be crucial for identifying deviations from normal operation. Anomalies in this context encompass not only outliers in individual data streams but also unusual combinations or patterns across multiple variables. Smart manufacturing systems, through the utilization of MVTS, achieve real-time monitoring and adaptive control. This involves advanced techniques in machine learning and data analytics, which play a pivotal role in extracting valuable insights from intricate data. These insights ultimately enhance efficiency and system reliability, reduce waste, and improve product quality. A comprehensive understanding of univariate and multivariate time series is an essential requirement to address challenges related to this data. In the realm of smart manufacturing data pre-processing, effective segmentation of time series data is a key challenge. Dividing time series into meaningful segments is vital for training accurate and efficient models, enabling pattern recognition, and facilitating anomaly identification. the forthcoming section will delve into the methodologies and significance of time series segmentation, particularly within the context of smart manufacturing.

3 Segmentation and Windowing in Time Series Analysis

Time series segmentation is a pivotal process in temporal data mining and signal processing, particularly for applications such as anomaly detection. It involves dividing a time series into segments according to specific criteria, simplifying and structuring the data for tasks like classification, pattern recognition, and anomaly detection. Depending on the application, the goals of segmentation can be quite diverse, essentially falling into various categories. These goals might include but are not limited to, enhancing the accuracy of predictive models, improving the efficiency of data processing, identifying critical events or changes in patterns, and facilitating a more nuanced understanding of the underlying processes within the data. Each application brings its own set of requirements and challenges to time series segmentation, underlining the importance of a tailored approach that aligns with the specific objectives and constraints of the task at hand. Below we will focus on two main types, each with unique goals and methodologies.

3.1 Segmentation for Time Series Reconstruction and Representation

- **Objective**: This segmentation aims to represent a time series through linear or non-linear approximations, focusing on minimizing the approximation error.
- **Methods**:
  - *Off-line Segmentation*: Algorithms with a global view of the data, potentially yielding better results due to comprehensive analysis. Examples include Top-Down, Bottom-Up [10], and k-Segmentation algorithms [11].

- *On-line Segmentation*: Suitable for real-time applications, making decisions with a local view of the data. The Sliding Window and Bottom-Up (SWAB) algorithm is a common example [12].

The evaluation of the model's performance using this segmentation approach often relies on assessing the least-squares reconstruction error of the approximated segments.

3.2 Segmentation at Characteristic Points of the Time Series

- **Objective:** Focusing on segmenting the time series at points of significant characteristic changes. Includes applications like efficient indexing and identifying perceptually important points (PIPs) [13].
- **Methods:** Techniques like the Sliding Window algorithm, using systems of orthogonal polynomials, are common. They employ various error criteria for determining segmentation points [14].

However, traditional measures like reconstruction error are not always ideal and depend heavily on the approximation algorithm. There is a need for metrics focusing on segmentation quality relative to user-defined points. An evident challenge stems from the absence of optimal evaluation metrics, impeding the precise quantification of segmentation results based on user-defined criteria. Addressing this research gap becomes imperative for the progression of the field. Since several segmentation techniques have been proposed, it is difficult to select a suitable one for a specific application if we do not know how to evaluate the performance of various segmentation methods. Several studies in this domain, as exemplified by the works of [15, 16], focus on evaluating segmentation quality by assessing the alignment of segmented points with user-designated significant points.

3.3 Locality in Time Series Analysis

An important consideration in MTVS analysis is the concept of *locality*. Locality suggests that temporally proximate data points are more likely to be correlated or exhibit similar patterns than those further apart. In practical terms, manufacturing processes implies data collected during the same shift or from the same production line are inherently more interrelated. Recognizing this concept is vital for anomaly detection, as it aids in identifying specific patterns and trends relevant to certain time

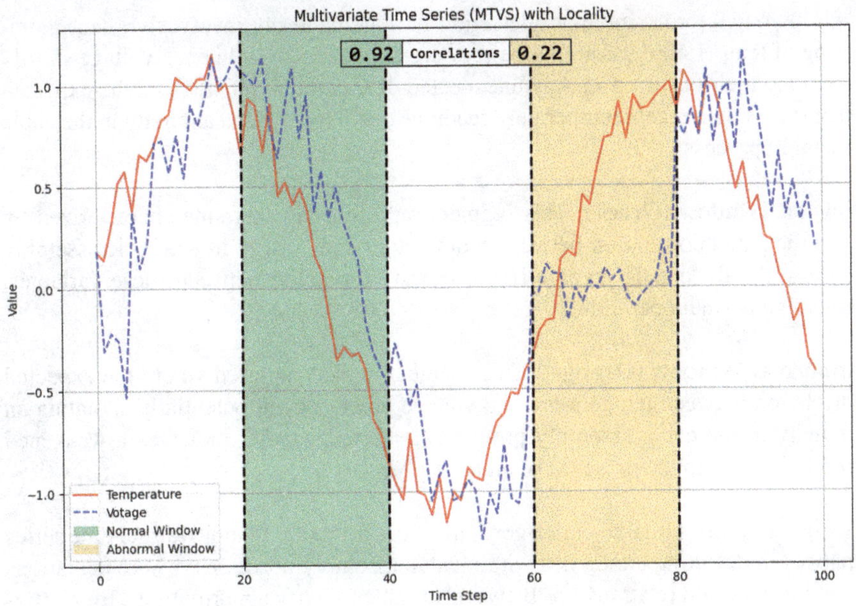

Fig. 4 Multivariate Time Series (MTVS) with locality

frames or operational conditions, which might be overlooked in more aggregated data analyses.

Figure 4, illustrates the concept of locality within multivariate time series (MVTS). **Lagging Correlation**: In this context, the plot demonstrates a temporal delay between certain parameters, such as temperature (depicted by the red line) and voltage (illustrated by the blue dashed line) data in a cable assembly process. Voltage fluctuations tend to follow changes in temperature with a delay, suggesting a localized relationship. This delay implies that temperature changes may precede corresponding alterations in voltage, highlighting the temporal correlation between these variables.

Temporal Proximity: When examining a narrow time window within the cable assembly process, temperature and voltage data within that window exhibit higher similarity compared to data outside of it. Peaks or troughs in temperature may correspond to analogous patterns in voltage shortly afterward, reflecting the proximity in time between these variables' fluctuations.

Pattern Recognition: The plot showcases a consistent pattern wherein changes in temperature are mirrored in voltage after a short interval within the cable assembly process. This consistent pattern suggests a temporal association between consecutive data points, which can be leveraged for anomaly detection.

In practical terms, anomaly detection in manufacturing involves vigilant monitoring of temperature variations, as they are anticipated to influence voltage shortly following their occurrence. Any unexpected deviation in voltage from the expected change based on recent temperature readings could indicate an anomaly in the cable assembly process [17].

Normal Window (Green): This segment represents a timeframe characterized by the anticipated correlation between temperature and voltage in the cable assembly process. A tall 'candle' symbolizes a robust correlation between these variables, indicating normal operation.

Abnormal Window (Orange): This segment depicts a period where the expected correlation between temperature and voltage breaks down, potentially signaling an anomaly in the cable assembly process. A shorter 'candle' indicates a weakened correlation due to the presence of an anomaly.

Figure 4 representation underscores the intricate nature of multivariate time series analysis in detecting anomalies within cable assembly processes, where deviations from the expected relationships between variables signify abnormalities. Emphasizing the importance of locality within smart manufacturing environments, it becomes evident that understanding temporal dependencies is crucial for effective anomaly detection and maintenance strategies. The acknowledgment of temporal locality within time series data profoundly influences the choice of analytical methodologies. While conventional approaches such as RNN networks, LSTM models, and Autoencoders excel in capturing long-term trends by smoothing short-term fluctuations, their ability to address complex inter-variable dependencies may be insufficient. This limitation could lead to overlooking crucial relationships within the data, thereby reducing their effectiveness in capturing nuanced patterns.

Here, the Transformer architecture [18–21] stands out-a robust framework adept at capturing both short and long-term dependencies, contextual nuances, and local patterns within the data. With the Transformer's ability to attend to relevant temporal contexts and relationships across various time series datasets, it becomes a potent tool for deciphering intricate patterns that might elude simpler models. By leveraging the Transformer's capacity to capture both local and global dependencies simultaneously, manufacturers can gain deeper insights into their processes, enabling proactive quality control and predictive maintenance. In essence, harnessing the power of the Transformer architecture allows manufacturers to navigate the complexity of multivariate time series data more effectively, leading to improved operational efficiency, reduced waste, and enhanced product quality in smart manufacturing environments. However, it is essential to carefully design its attention mechanism to effectively capture these dependencies, as we will elaborate on when presenting the proposed methodologies in upcoming sections.

In summary, the exploration of time series data unveils the complexity and crucial role of segmentation, laying the groundwork for a variety of time series mining tasks. These tasks encompass dimension reduction, representation, classification, clustering, prediction, and anomaly detection. In this context, Multivariate Time Series (MVTS) data presents unique challenges. This complexity is due to high-order dependencies and the essential need to capture data locality for effective analysis. MVTS data is defined by the interdependencies among multiple variables and the temporal nature of these relationships, distinguishing it from simpler time series datasets. Traditional methods for anomaly detection often fall short when applied to MVTS data. This is because anomalies can manifest not only in individual variables but also in the dynamics of relationships between variables over time. Thus, there is a pronounced need for advanced methods that are adept at capturing both temporal dependencies and the localized nature of anomalies. Given this backdrop, it becomes increasingly important to delve deeper into the anomaly detection domain within time series data, with a special emphasis on the challenges and methods pertinent to MVTS. This focus is crucial for developing more sophisticated and effective anomaly detection techniques that can navigate the intricacies of multivariate relationships and temporal dependencies.

4 MVTS Anomaly Detection in Smart Manufacturing

As we embark on a deeper exploration of anomaly detection, it becomes crucial to center our attention on comprehending the intricacies of anomaly detection within the context of Multivariate Time Series (MVTS) data. This focus is imperative due to the distinctive challenges posed by MVTS data, particularly within the domain of smart manufacturing (SM). Furthermore, anomaly detection is compounded by the intricacies inherent in MVTS data in SM, where the complexity escalates with the presence of noise and the dynamic nature of manufacturing processes. This discernment is pivotal as it entails the recognition of subtle variations amidst fluctuating data patterns.

When we approach the mathematical formulation of anomaly detection in MVTS data, we encounter notations such as $m_{i1}, m_{i2}, \ldots, m_{ip}$. Here, the indices i and p are crucial for our analysis:

- The subscript i indicates the time index at which observations are made, essential for tracking the temporal sequence of data. This index reflects the specific moment (e.g., the ith minute, hour, or day) when measurements were taken. Within the context of smart manufacturing, i enables us to identify the exact timing of potential anomalies, allowing for prompt intervention.
- The subscript p denotes the number of variables monitored in the MVTS. Each m_{ij} (where $j = 1, 2, \ldots, p$) corresponds to the measurement of the jth variable at time i. This multivariate approach distinguishes MVTS from univariate time series, offering a holistic view of how various factors (such as temperature, pressure, vibration, or speed) interact over time.

For anomaly detection, we consider two primary mathematical approaches:

1. **Deviation from Expected Behavior**: This method focuses on identifying data points $m_{i1}, m_{i2}, \ldots, m_{ip}$ that significantly deviate from the expected behavior, represented by $E[f(M)]$. Here, $E[f(M)]$ symbolizes the conventional or normal behavior of the system, as inferred from historical data. Anomalies are detected when the function f, applied to the observed data at time i across all p variables, does not closely approximate the expected value, signaling a potential issue or abnormality in the manufacturing process.

$$f(m_{i1}, m_{i2}, ..., m_{ip}) \not\approx E[f(M)] \tag{6}$$

M, corresponds to the matrix representation of Multivariate Time Series (MVTS) data, as previously elucidated in Sect. 2 (Eq. 4).

2. **Function Modeling of Normal Behavior**: The second formulation involves using a mathematical function $f(x_t, \Theta)$ to model the normal behavior of the MVTS data.

$$A_t = f(x_t, \Theta) \tag{7}$$

Here, A_t represents the anomaly score at time t, which measures how anomalous the data is compared to what the model predicts. The parameter Θ represents the set of model parameters that the function relies on for its predictions.

In addition to these formulations, understanding the temporal correlations within MVTS data is crucial. This involves recognizing how the values of multiple variables at a specific time X_t are related not only to their past values but also to the past values of other variables in the system. This complex interdependency can be mathematically represented using a Multivariate Vector Autoregressive (VAR) model:

$$X_t = \Phi_1 X_{t-1} + \Phi_2 X_{t-2} + \cdots + \Phi_p X_{t-p} + \epsilon_t \tag{8}$$

where:

- X_t is a vector representing the values of all variables at time t.
- $\Phi_1, \Phi_2, \ldots, \Phi_p$ are coefficient matrices that capture the interdependencies between the current value of the variables and their values at previous time steps. Each Φ_i corresponds to a specific time lag i, indicating the strength and direction of the relationship across variables.
- p represents the number of time lags considered, reflecting how far back in time the model looks to understand the correlations.
- ϵ_t is a vector of error terms, accounting for any unexplained variance or randomness in the data at time t.

This model allows for a comprehensive analysis of how each variable within the MVTS data influences and is influenced by, the others over time, providing a more complete understanding of the system's dynamics. By using these mathematical formulations, we can analyze MVTS data to detect anomalies by either comparing observed data to expected behavior or by modeling the normal behavior of the data and identifying deviations from the model's predictions. Additionally, incorporating temporal correlations through VAR models allows us to capture dependencies between variables across different time points, enhancing our ability to detect anomalies in MVTS data. Such modeling is essential for identifying and understanding the intricate patterns and potential irregularities in production.

In Sect. 2, we have undertaken a comprehensive exploration of the components of Multivariate Time Series (MVTS) data within the context of smart manufacturing. We have delved into the intricacies of trend analysis, seasonal adjustments, cyclical component understanding, and the challenges of handling random variations in time series data. This deep dive has provided us with a fundamental understanding of the complexities inherent in manufacturing data.

4.1 Time Series Analysis

Armed with this profound knowledge of time series components and their significance in the manufacturing context, let's further investigate how the analysis of these components plays a pivotal role in anomaly detection (AD). The ability to dissect and comprehend these components becomes increasingly important as we explore the applications and nuances of decomposition-based AD techniques. These techniques harness the power of time series component analysis to detect anomalies effectively.

Trend Analysis: (i) *Univariate*: Identifying deviations from the trend in a univariate setting can signal potential anomalies, such as a gradual increase in machine temperature indicating wear and tear. (ii) *MVTS*: Trend analysis in MVTS is more intricate but provides richer insights. Anomalies may be detected when the combined trend of multiple variables deviates from the expected pattern, indicating changes in the manufacturing process [22, 23].

Seasonal Adjustments: (i) *Univariate*: Recognizing and adjusting for seasonal patterns helps differentiate between normal cyclical changes and true anomalies. (ii) *MVTS:* In MVTS, understanding the interplay between seasonal patterns of different variables is key. Anomalies may be indicated by deviations from expected synchronized patterns across variables [23].

Cyclical Component Understanding: (i) *Univariate*: Acknowledging cyclical changes, distinct from seasonal patterns, aids in identifying anomalies related to longer-term oscillations in data. (ii) *MVTS*: Cyclical components in a multivariate context can be complex. Anomalies might be indicated by unusual alignments or misalignments of these cycles across different variables [24].

Handling Random Variations: (i) *Univariate*: Distinguishing random noise from actual anomalies is essential. (ii) *MVTS*: In MVTS, the challenge lies in determining whether random variations in one variable correlate with changes in others, suggesting systemic issues rather than random noise [24].

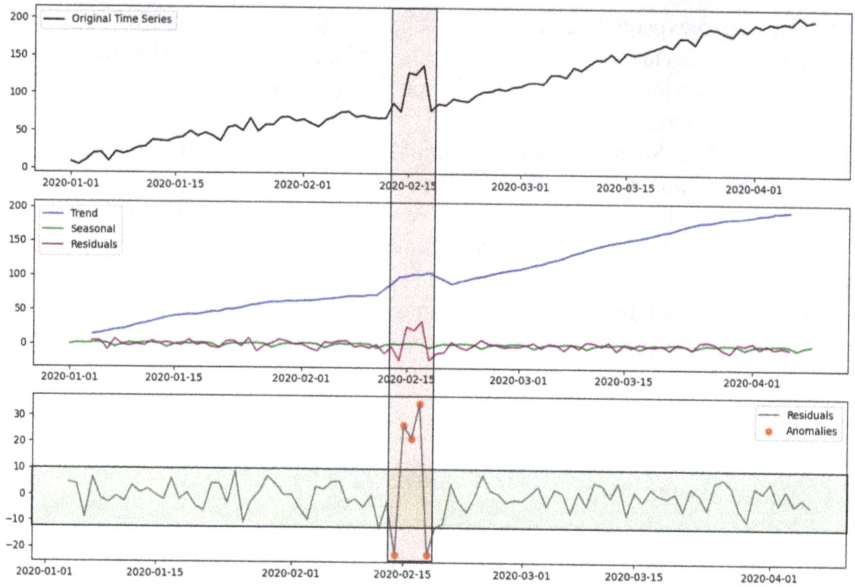

Fig. 5 Identification of anomalies through decomposition in time series analysis

Building upon the detailed exploration of trend analysis, seasonal adjustments, cyclical component understanding, and handling random variations in time series, it becomes evident that Decomposition-based Anomaly Detection (AD) in Time Series (TS) offers a strategic and effective approach for dissecting complex datasets. This method dissects the data into fundamental elements, focusing on each separately, with particular attention to the residuals [25]. Residuals represent the portion of data remaining after removing established trends and seasonal variations, and they are crucial for spotting irregularities. A notable deviation in the residuals, incongruent with the predicted trend or seasonality, signals a potential anomaly. This method's emphasis on residuals improves detection precision by separating atypical occurrences from normal patterns, look at Fig. 5.

A detailed understanding of time series components and the decomposition approach is essential for grasping complex AD techniques.

In the context of anomaly detection in smart manufacturing, the concrete definition of an anomaly differs in the literature. Given the ordering characteristic of time series data, anomalous data points have been classified into point, sequence, and contextual anomalies to consider length and context information in the classification. When subdividing a time series into fixed-size subsequences (e.g., via windowing), the normality of individual points can be defined by their distance to (nearest) neighbor patterns w.r.t. certain similarity and distance measures. This leads to the anomaly concept of discords. This approach's significance becomes even more pronounced when considering its application in various aspects of anomaly detection. By integrating these methodologies, manufacturers can craft advanced preventative maintenance strategies, leading to the overall enhancement of productivity and reliability within manufacturing systems. Thus, it is clear that anomaly detection extends beyond mere identification of data irregularities; it encompasses a predictive capability essential for optimizing system efficiency and safety. This predictive dimension sets it apart from traditional outlier detection methods, as it focuses not only on recognizing data irregularities but also on anticipating potential disruptions that align with the specific operational dynamics and safety requirements of the manufacturing environment.

Understanding the concept of anomaly detection in smart manufacturing, especially for quality inspection, requires a profound comprehension of the various anomaly types present in manufacturing data. These anomalies serve as signals indicating potential deviations from established operational norms, which could result in quality issues, reduced production efficiency, or machinery breakdowns. The core anomaly categories in this context include point anomalies, contextual anomalies, collective anomalies, and sequential anomalies. Recognizing the distinct deviation patterns associated with each type is essential for implementing effective quality control measures and maintaining the operational excellence of smart manufacturing systems.

4.2 *Types of Anomalies*

As we delve deeper into this topic, we will explore the different types of anomalies and how they impact the quality and efficiency of smart manufacturing processes.

1. **Point Anomaly**: A point anomaly is a single data point that significantly diverges from the norm Fig. 6. In a manufacturing setting, this might represent an unexpected event like a sudden spike in machine temperature or an irregularity in material quality. Though they are relatively straightforward to detect, understanding their causes and impact on the manufacturing process is crucial [26].

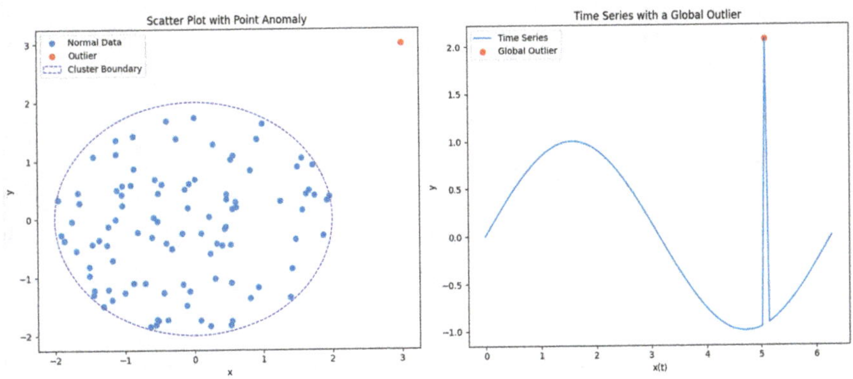

Fig. 6 Point anomaly

Below is the Python code to plot the Fig. 6, full code can be found in this GitHub repository [27]

```
import matplotlib.pyplot as plt
import numpy as np

# Plot 1: Scatter plot with a point anomaly
# read normal data points
x_scatter = np.loadtxt('x_scatter.txt')
y_scatter = np.loadtxt('y_scatter.txt')

# Create a point anomaly
# x_scatter = np.append(x_scatter, 3)
# y_scatter = np.append(y_scatter, 3)

# # Read point anomaly

x_anomaly_outside = np.loadtxt('x_anomaly_outside.txt')
y_anomaly_outside = np.loadtxt('y_anomaly_outside.txt')

plt.figure(figsize=(14, 6))

plt.subplot(1, 2, 1)
plt.scatter(x_scatter, y_scatter,color='black',\
label='Normal Data')
# Plot the point anomalies outside the main cluster
plt.scatter(x_anomaly_outside, y_anomaly_outside, \
color='red', label='Outlier')

# Draw a circle around the collective anomalies
plt.gca().add_patch(plt.Circle((2, 2), 0.3, color='red',\
fill=False, linestyle='--', linewidth=1.5))


for i in range(len(x_anomaly_outside)):
    plt.annotate('Outliers', (x_anomaly_outside[i],\
    y_anomaly_outside[i]), textcoords="offset points",\
    xytext=(-15,10), ha='center',\
    arrowprops=dict(arrowstyle='->',\
    connectionstyle='arc3'))

plt.xlabel('x')
plt.ylabel('y')
plt.title('Scatter Plot with a Point Anomaly')
plt.legend()
plt.grid(True)

# Plot 2: Time series plot with a global outlier
t = np.linspace(0, 2*np.pi, 100)
x_time_series = np.sin(t)

# Introduce a global outlier
x_time_series[80] += 3
```

```
plt.subplot(1, 2, 2)
plt.plot(t, x_time_series, label='Time Series', \
color='black')
plt.scatter(t[80], x_time_series[80], color='red', \
label='Global Outlier')

plt.xlabel('x(t)')
plt.ylabel('y')
plt.title('Time Series with a Global Outlier')
plt.legend()
# add grid
plt.grid(True)

# Show the plots
plt.tight_layout()
plt.show()
```

2. **Contextual Anomaly**: Contextual anomalies, or conditional outliers, are data points that deviate significantly from others in the same context. For example, a change in machine performance under certain conditions or unusual production output during specific shifts (Fig. 7). Recognizing these anomalies requires understanding the specific context, such as time, environmental conditions, or operational settings [28].
3. **Collective Anomaly**: Collective anomalies involve a group of data points that collectively indicate an abnormal pattern (Fig. 8). In smart manufacturing, these could be minor but consistent fluctuations in machine performance indicating a potential equipment failure or a gradual decline in product quality over time. Detecting such anomalies is key to preventing more significant issues [29].
4. **Sequential Anomaly**: Specific to time-series data or sequential processes, sequential anomalies occur when a series of events or data points deviate from the typical pattern (Fig. 9). An example in manufacturing could be a sequence of machine operations occurring out of their usual order, possibly signaling a problem in the automated process [30].

Each type of anomaly requires different methods of detection and analysis, and understanding these differences is crucial for effective quality inspection in smart manufacturing environments. With the integration of machine learning in smart manufacturing, the approach to anomaly detection has significantly evolved. Machine learning allows for automated analysis of vast quantities of real-time data, enhancing the ability to efficiently detect and deeply understand the nature of anomalies. Techniques such as unsupervised, supervised, and semi-supervised learning are employed for different types of anomalies, providing a comprehensive strategy for maintaining quality and efficiency in manufacturing processes. While the advancement in anomaly detection (AD) techniques has significantly enhanced our ability to detect and analyze anomalies in smart manufacturing, this progress is accompanied by a set of challenges. It is essential to explore the historical evolution of AD methods to understand the state-of-the-art (SOA) and their relevance in the context of complex

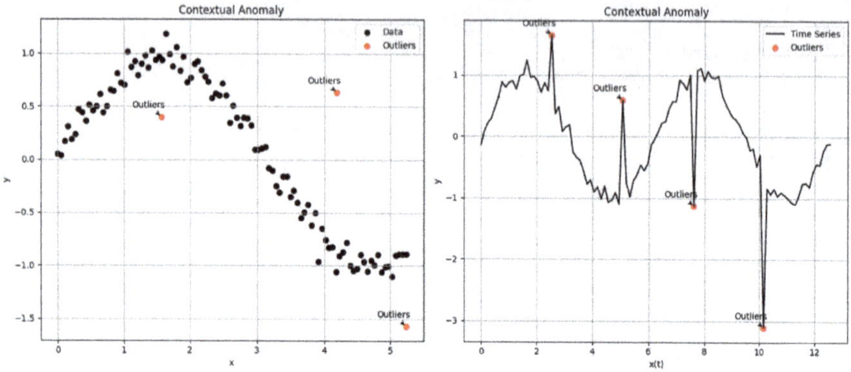

Fig. 7 Contextual anomaly

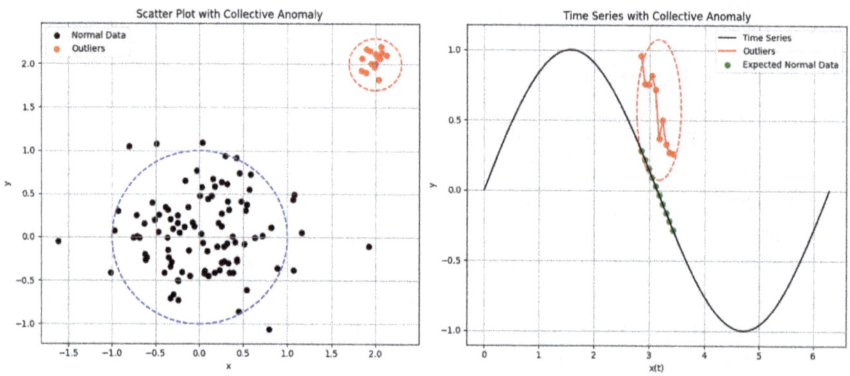

Fig. 8 Collective anomaly

and dynamic environments like smart manufacturing. This section aims to provide a comprehensive review of how AD techniques have developed over time, moving from traditional methods to the sophisticated approaches employed in the eras of Industry 4.0 and 5.0. By examining the academic evolution of AD and highlighting the gap between theoretical advancements and their practical application, we can appreciate the innovative solutions that have emerged to address the unique challenges faced by smart manufacturing systems. This literature overview not only aids in understanding the current capabilities and limitations of AD in smart manufacturing but also sets the stage for anticipating future innovations and directions in this continually evolving field.

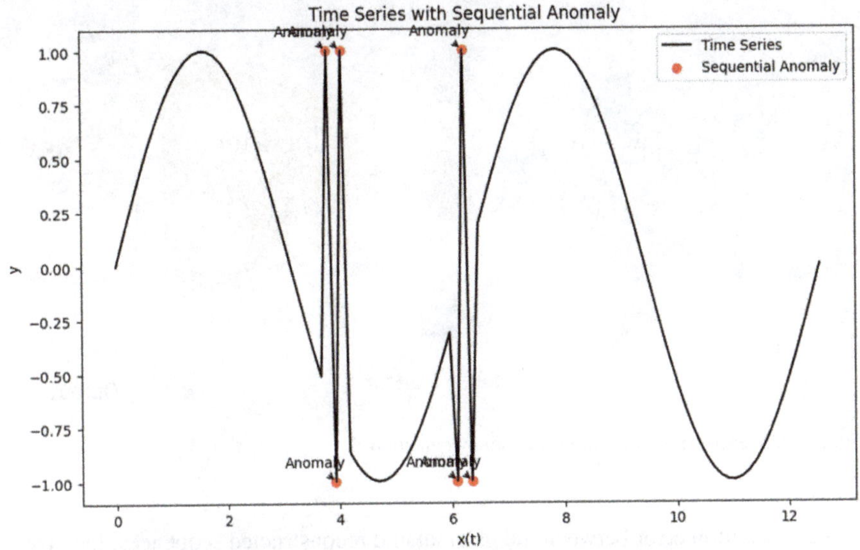

Fig. 9 Sequential anomaly

5 A Step by Step Anomaly Detection on Industrial Data

In this section, we provide a comprehensive example that demonstrates how to detect anomalies in sequence or time series data using an LSTM Autoencoder.

An autoencoder is a type of neural network that is trained to replicate its input by transforming the input into a lower-dimensional space (the encoding step) and then reconstructing the input from this lower-dimensional representation (the decoding step). Training an autoencoder does not require labeled data.

While an autoencoder itself does not directly detect anomalies, it can be used as a tool for anomaly detection. By training an autoencoder using only representative (normal) data, we create a model that can accurately reconstruct its input data using the features learned from the normal data, Fig. 10.

Here is the process to detect anomalies using an autoencoder:

1. **Training the Autoencoder**: Train the autoencoder using a dataset that only contains normal (non-anomalous) time series data. This ensures that the model learns to reconstruct data that is representative of normal conditions.
2. **Reconstruction Error**: To detect anomalies, input the observation (time series data) into the trained autoencoder and measure the error between the original observation and the reconstructed observation. The reconstruction error is typically calculated using a metric such as Mean Squared Error (MSE).
3. **Anomaly Detection**: A large reconstruction error indicates that the original observation contains features that are unrepresentative of the data used to train the autoencoder, suggesting that it is anomalous. By examining the element-wise

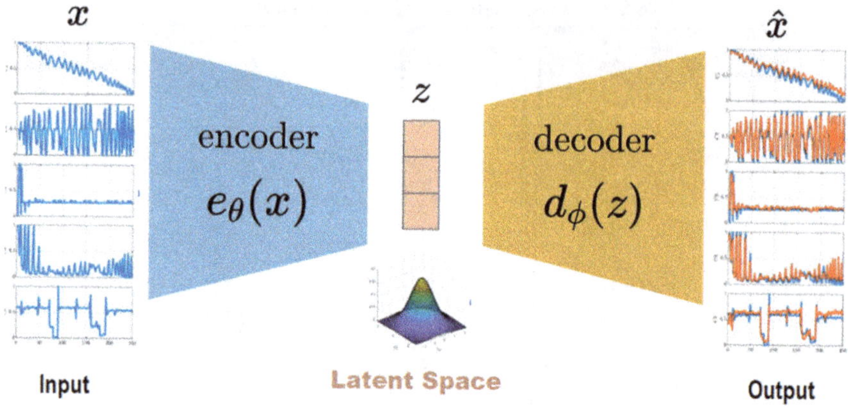

Fig. 10 Autoencoder architecture for anomaly detection

reconstruction error between the original and reconstructed sequences, localized regions of anomalies can be identified.

Here's a step-by-step example:

1. **Data Preparation**:
   - Collect a dataset of normal time series data.
   - Preprocess the data (e.g., normalization, scaling).
2. **Train the Autoencoder**:
   - Split the normal data into training and validation sets.
   - Define the autoencoder architecture (e.g., number of layers, neurons, activation functions).
   - Train the autoencoder on the normal data, minimizing the reconstruction error.
3. **Calculate Reconstruction Error**:
   - Input a new observation (which may contain anomalies) into the trained autoencoder.
   - Obtain the reconstructed observation from the autoencoder.
   - Calculate the reconstruction error for the observation.
4. **Identify Anomalies**:
   - Compare the reconstruction error against a predefined threshold. If the error exceeds the threshold, the observation is considered anomalous.
   - Analyze the element-wise reconstruction error to pinpoint localized regions of anomalies within the sequence.

Python Implementation: Here's a Python implementation of the above steps using Keras and Matplotlib:

```
mport numpy as np
import matplotlib.pyplot as plt
from sklearn.preprocessing import MinMaxScaler
from tensorflow.keras.models import Sequential
from tensorflow.keras.layers import Dense, LSTM, RepeatVector, \
TimeDistributed

# Generate synthetic normal data for training
def generate_data(n_samples, n_features, timesteps):
    data = np.random.normal(size=(n_samples, timesteps,\
    n_features))
    return data

# Data Preparation
data = generate_data(n_samples=1000, n_features=1, timesteps=\
50)
scaler = MinMaxScaler()
data_scaled = scaler.fit_transform(data.reshape(-1, 1)).\
reshape(data.shape)

# Train-Test Split
train_size = int(0.8 * data_scaled.shape[0])
train_data = data_scaled[:train_size]
test_data = data_scaled[train_size:]

# Autoencoder Model
model = Sequential([
    LSTM(64, activation='relu', input_shape=(train_data.\
    shape[1], train_data.shape[2]), return_sequences=False),
    RepeatVector(train_data.shape[1]),
    LSTM(64, activation='relu', return_sequences=True),
    TimeDistributed(Dense(train_data.shape[2]))
])

model.compile(optimizer='adam', loss='mse')
model.fit(train_data, train_data, epochs=50, batch_size=32,\
validation_split=0.1)

# Plot training & validation loss values
plt.figure(figsize=(10, 6))
plt.plot(history.history['loss'], label='Training loss')
plt.plot(history.history['val_loss'], label='Validation loss')
plt.title('Model loss')
plt.ylabel('Loss')
plt.xlabel('Epoch')
plt.legend(loc='upper right')
plt.show()

# Calculate Reconstruction Error
def calculate_reconstruction_error(model, data):
    reconstructed = model.predict(data)
```

```
    mse = np.mean(np.power(data - reconstructed, 2),\
    axis=(1, 2))
    return mse, reconstructed

mse_train, _ = calculate_reconstruction_error(model, \
train_data)
threshold = np.percentile(mse_train, 95) \
# Set threshold as 95th percentile of training reconstruction\
#error

# Test on new data (including anomalies)
test_data_with_anomalies = generate_data(n_samples=200, \
n_features=1, timesteps=50)
test_data_with_anomalies[10:20] += 2  # Inject anomalies
mse_test, reconstructed_test = calculate_reconstruction_error\
(model, test_data_with_anomalies)

# Identify Anomalies
anomalies = mse_test > threshold

# Plot Reconstruction Error
plt.figure(figsize=(10, 6))
plt.plot(mse_test, label='Reconstruction error')
plt.axhline(y=threshold, color='r', linestyle='--', \
label='Threshold')
plt.xlabel('Sample index')
plt.ylabel('MSE')
plt.title('Reconstruction Error with Anomalies')
plt.legend()
plt.show()

# Plot an example of anomaly detection
index = 15  # Index with an anomaly
plt.figure(figsize=(10, 6))
plt.plot(test_data_with_anomalies[index], label='Original')
plt.plot(reconstructed_test[index], label='Reconstructed')
plt.title(f'Original vs Reconstructed for Sample {index}')
plt.legend()
plt.show()
```

This implementation includes the following steps:

- Data generation and preprocessing.
- Autoencoder model definition and training.
- Calculation of reconstruction error and anomaly detection.
- Visualization of reconstruction error and anomaly detection results.

Figure 11 shows the model loss development over the training epochs and Fig. 12 shows a Box Plot of Reconstruction Errors.

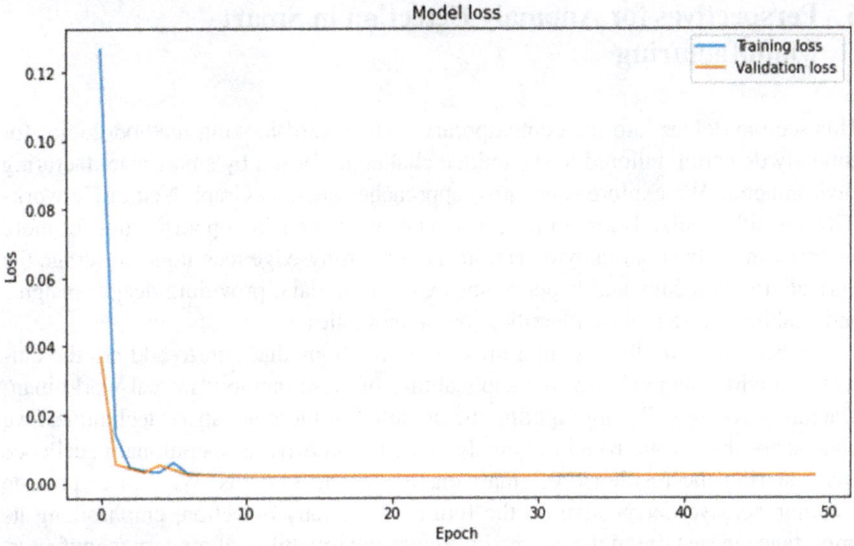

Fig. 11 Model loss

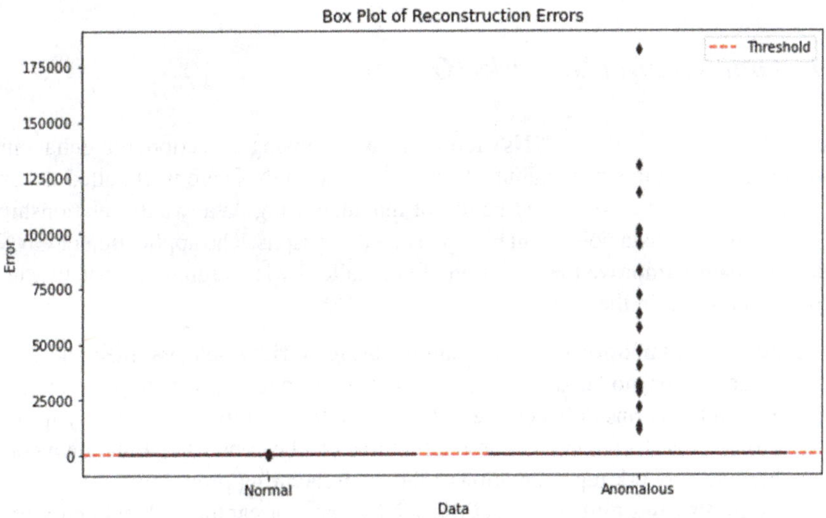

Fig. 12 Box plot of reconstruction errors

This approach allows for effective anomaly detection in time series data by leveraging the ability of autoencoders to learn and reconstruct normal patterns, making deviations from these patterns readily apparent through increased reconstruction error, [31].

6 Perspectives for Anomaly Detection in Smart Manufacturing

This section delves into the contemporary and forward-looking methodologies for anomaly detection tailored to the unique challenges posed by smart manufacturing environments. We explore innovative approaches such as Graph Neural Networks (GNNs) and Tabular Transformers, which present promising opportunities for more accurate and robust anomaly detection. These cutting-edge techniques leverage the intricate relationships and dependencies within the data, providing deeper insights and enabling more precise identification of anomalies.

Furthermore, we discuss future research directions that aim to address the current limitations and enhance the applicability of these methods in real-world manufacturing settings. By highlighting the potential of these advanced techniques, we underscore the pivotal role of anomaly detection in driving operational excellence and fostering the resilience of smart manufacturing systems. We aim to provide a comprehensive perspective on the future of anomaly detection, emphasizing its importance in sustaining the competitive edge and reliability of modern manufacturing processes.

6.1 Graph Neural Networks (GNNs)

Graph Neural Networks (GNNs) represent a promising direction for enhancing anomaly detection in smart manufacturing processes. GNNs are well-suited to handle the complex, interconnected nature of manufacturing data, where relationships between different data points can be represented as graphs. The application of GNNs can significantly improve the detection of anomalies by leveraging the structure and dependencies within the data.

1. **Graph Construction**: In smart manufacturing, various sensors, machines, and processes can be modeled as nodes in a graph, with edges representing the relationships or interactions between them. For example, machines that directly impact each other's performance in a production line can be connected, creating a comprehensive network representation of the manufacturing process.
2. **Message Passing and Aggregation**: GNNs excel at capturing local and global dependencies through message passing and aggregation mechanisms. In the context of anomaly detection, this allows the model to consider the influence of neighboring nodes, making it possible to detect anomalies that may not be apparent when examining individual nodes in isolation.
3. **Temporal GNNs**: Extending GNNs to handle temporal data (e.g., Temporal Graph Networks) can enhance their applicability to time series data commonly found in manufacturing. By incorporating temporal information, these models can capture the dynamic nature of manufacturing processes, identifying evolving anomalies.

4. **Anomaly Detection Algorithms**: GNN-based anomaly detection algorithms, such as Graph Convolutional Networks (GCNs) and Graph Attention Networks (GATs), can be tailored to identify both point anomalies and more complex patterns, such as sequential or collective anomalies, by leveraging the rich contextual information available in the graph structure.

6.2 Tabular Transformers

Transformers, originally developed for natural language processing, have demonstrated exceptional capabilities in handling sequential data. Recent advancements have adapted transformers for tabular data, making them highly relevant for anomaly detection in smart manufacturing [32].

1. **Sequential Data Handling**: Tabular transformers can efficiently process multivariate time series data, capturing complex dependencies across different features and time steps. This is particularly beneficial for detecting anomalies that involve intricate patterns across multiple sensor readings.
2. **Attention Mechanism**: The attention mechanism inherent in transformers allows the model to focus on the most relevant parts of the data, improving the identification of subtle anomalies. For example, in a scenario where specific sensor readings exhibit anomalous behavior only under certain conditions, the attention mechanism can highlight these critical interactions.
3. **Scalability and Flexibility**: Transformers are highly scalable and can be trained on large datasets, making them suitable for the vast amounts of data generated in smart manufacturing environments. Their flexibility allows for the incorporation of domain knowledge and the customization of model architectures to suit specific applications better.
4. **Hybrid Approaches**: Combining transformers with other machine learning techniques, such as convolutional neural networks (CNNs) or recurrent neural networks (RNNs), can further enhance their performance. These hybrid models can leverage the strengths of different architectures, providing a more robust solution for anomaly detection.

6.3 Future Research Directions

1. **Integration with IIoT**: Leveraging the Industrial Internet of Things (IIoT) to gather real-time data from various sources and feeding this data into advanced anomaly detection models can significantly enhance the timeliness and accuracy of anomaly detection.
2. **Explainable AI (XAI)**: As the complexity of models like GNNs and transformers increases, the need for explainable AI becomes critical. Developing methods to

interpret and explain the decisions made by these models will be essential for gaining trust and ensuring compliance with industry standards.

3. **Federated Learning**: Implementing federated learning approaches can enable the training of anomaly detection models across multiple manufacturing sites without the need to centralize data. This can enhance data privacy and security while improving model generalization.
4. **Adaptive Models**: Developing adaptive models that can continuously learn from new data and adapt to changes in manufacturing processes will be crucial for maintaining high detection performance over time.

In summary, the integration of advanced techniques such as Graph Neural Networks and Tabular Transformers holds significant promise for the future of anomaly detection in smart manufacturing. These approaches can address the complex and dynamic nature of manufacturing data, providing more accurate and timely identification of anomalies, and ultimately driving operational excellence and innovation in the manufacturing industry.

7 Conclusion

This chapter has examined the evolving landscape of anomaly detection within the context of smart manufacturing, emphasizing the integration of advanced technologies such as IIoT, AI, and machine learning. As smart manufacturing environments become increasingly complex and data-driven, the need for robust anomaly detection methods has become more pronounced. We have covered the importance of leveraging multivariate time series data, highlighting the critical role of segmentation, windowing, and effective data representation techniques in enhancing anomaly detection capabilities.

We differentiated between various types of anomalies-contextual, point, sequential, and collective-each presenting unique challenges and requiring distinct approaches for detection. This nuanced understanding of anomaly types is essential for developing tailored anomaly detection strategies that can more accurately identify and address the specific deviations that occur in manufacturing processes.

The exploration of characteristic points and locality in time series analysis has underscored the necessity of sophisticated approaches that cater to the specific dynamics of manufacturing processes. By focusing on these methods, manufacturers can identify and address anomalies more swiftly and accurately, thereby preventing costly downtime and enhancing production quality.

Furthermore, the historical overview of anomaly detection techniques provided in this chapter demonstrates the rapid advancements in this field and sets the stage for future innovations. Looking ahead, the integration of more sophisticated AI models and the increasing automation of data analysis processes promise to further refine anomaly detection strategies in smart manufacturing.

The diverse applications discussed herein reveal that anomaly detection is not only crucial for maintaining operational efficiency but is also pivotal in driving the industry toward unprecedented levels of reliability and precision. By continuing to develop and implement advanced anomaly detection methods, the manufacturing sector can better adapt to the challenges of Industry 5.0, ensuring that it remains competitive in a rapidly evolving technological landscape.

Thus, the journey toward operational excellence in smart manufacturing is significantly catalyzed by advances in anomaly detection. Researchers and practitioners alike must foster continuous improvement in these technologies, ensuring they keep pace with the complexity and demands of modern manufacturing processes.

References

1. Dan Luo ST, Dolgui A (2023) A state-of-the-art on production planning in industry 4.0. Int J Prod Res 61:6602–6632
2. Maddikunta PKR, Pham Q-V, Deepa PBN, Dev K, Gadekallu TR, Ruby R, Liyanage M (2022) Industry 5.0: a survey on enabling technologies and potential applications. J Ind Inf Int 26:100257
3. Wang B, Tao F, Fang X, Liu C, Liu Y, Freiheit T (2021) Smart manufacturing and intelligent manufacturing: a comparative review. Engineering 7:738–757
4. Resek RW (1966) Investment by manufacturing firms: a quarterly time series analysis of industry data. Rev. Econ. Stat. 48:322–333
5. Zarnowitz V, Ozyildirim A (2006) Time series decomposition and measurement of business cycles, trends and growth cycles. J Mon Econ 53:1717–1739
6. Kusiak A (2000) Decomposition in data mining: an industrial case study. IEEE Trans Electron Pack Manuf 23:345–353
7. De Santo A, Ferraro A, Galli A, Moscato V, Sperlì G (2022) Evaluating time series encoding techniques for predictive maintenance. Expert Syst Appl 210:118435
8. Khalid A, Javaid N, Guizani M, Alhussein M, Aurangzeb K, Ilahi M (2018) Towards dynamic coordination among home appliances using multi-objective energy optimization for demand side management in smart buildings. IEEE Access 6:19509–19529
9. Kapp V, May MC, Lanza G, Wuest T (2020) Pattern recognition in multivariate time series: Towards an automated event detection method for smart manufacturing systems. J Manuf Mater Proc 4
10. Keogh E, Chu S, Hart D, Pazzani M (2004) Segmenting time series: a survey and novel approach. Data Mining Time Ser Databases 57:1–22
11. Bellman R (1961) On the approximation of curves by line segments using dynamic programming. Commun ACM 4:284
12. Keogh E, Chu S, Hart D, Pazzani M (2001) An online algorithm for segmenting time series. In: Proceedings 2001 IEEE international conference on data mining, pp 289–296
13. Chung F, Fu T-C, Luk R, Ng V (2001) Flexible time series pattern matching based on perceptually important points. In: International joint conference on artificial intelligence workshop on learning from temporal and spatial data 1–7 (2001)
14. Gensler A, Gruber T, Sick B (2013) Blazing fast time series segmentation based on update techniques for polynomial approximations. In: 2013 IEEE 13th international conference on data mining workshops, pp 1002–1011
15. Gensler A, Sick B (2014) Novel criteria to measure performance of time series segmentation techniques. CEUR Workshop Proc 1226:193–204

16. Thuy HTT, Anh DT, Chau VTN (2017) Comparing three time series segmentation methods via novel evaluation criteria. In: 2017 2nd international conferences on information technology, information systems and electrical engineering (ICITISEE), pp 171–176
17. Benkabou S-E, Benabdeslem K, Kraus V, Bourhis K, Canitia B (2022) Local anomaly detection for multivariate time series by temporal dependency based on poisson model. IEEE Trans Neural Netw Learn Syst 33:6701–6711
18. Tuli S, Casale G, Jennings NR (2022) Tranad: deep transformer networks for anomaly detection in multivariate time series data. Proc VLDB Endow 15:1201–1214
19. Xu J, Wu H, Wang J, Long M (2021) Anomaly transformer: Time series anomaly detection with association discrepancy. ArXiv:abs/2110.02642
20. Chen Z, Chen D, Zhang X, Yuan Z, Cheng X (2022) Learning graph structures with transformer for multivariate time-series anomaly detection in IoT. IEEE Int Things J 9:9179–9189
21. Li B, Cui W, Wang W, Zhang L, Chen Z, Wu M (2021) Two-stream convolution augmented transformer for human activity recognition. In: Proceedings of the AAAI conference on artificial intelligence, vol 35, pp 286–293
22. Killick PFR, Eckley IA (2012) Optimal detection of changepoints with a linear computational cost. J Am Stat Assoc 107:1590–1598
23. Zhang Z, Wang R, Ding R, Gu Y (2024) Unravel anomalies: An end-to-end seasonal-trend decomposition approach for time series anomaly detection. In: 2024 IEEE international conference on acoustics, speech and signal processing (ICASSP), pp 301–306. ArXiv:2310.00268v2 [cs.LG]
24. Mishra A, Sriharsha R, Zhong S (2021) Onlinestl: scaling time series decomposition by 100x. arXiv:2107.09110. ArXiv:2107.09110v4 [cs.LG]
25. Zhang C, Zhou T, Wen Q, Sun L (2022) Tfad: a decomposition time series anomaly detection architecture with time-frequency analysis. In: Proceedings of the 31st ACM international conference on information & knowledge management, CIKM '22, Association for Computing Machinery, New York, NY, USA, pp 2497–2507
26. Torkamani S, Lohweg V (2017) Survey on time series motif discovery. Wiley Interdiscip Rev: Data Mining Knowl Discov 7:1199
27. Orabi M (2024) Mtvs ploting. https://github.com/mousaab/MTVS_Ploting. GitHub
28. Mueen A (2014) Time series motif discovery: dimensions and applications. Wiley Interdiscip Rev: Data Mining Knowl Discov 4:152–159
29. Du X, Jin R, Ding L, Lee VE, Thornton JH (2009) Migration motif: a spatial-temporal pattern mining approach for financial markets. In: Elder J, Fogelman FS, Flach P, Zaki M (eds), Proceedings of the 15th ACM SIGKDD international conference on Knowledge discovery and data mining—KDD '09. ACM Press, New York, USA, pp 1135–1144
30. Pevzner PA, Sze S-H (2000) Combinatorial approaches to finding subtle signals in DNA sequences. In: Bourne PE, Gribskov M, Altman RB, Jensen N, Hope DA, Lengauer T, Mitchell JC, Scheeff ED, Smith C, Strande S, Weissig H (eds), Proceedings of the 8th international conference on intelligent systems for molecular biology. AAAI Press, Palo Alto, USA, pp 269–278
31. Orabi M (2023) Anomaly detection in manufacturing industry, https://medium.com/@moussab.orabi/anomaly-detection-in-manufacturing-industry-2400c0c9da9d. Accessed 17 Nov 2023
32. Huang X, Khetan A, Cvitkovic M, Karnin Z (2020) Tabtransformer: Tabular data modeling using contextual embeddings. arXiv:2012.06678

Human-Centered Edge Artificial Intelligence for Smart Factory Applications in Industry 5.0: A Review and Perspective

Le Hoang Nguyen, Kim Duc Tran, Xianyi Zeng, and Kim Phuc Tran

Abstract The integration of Human-Centered Edge Artificial Intelligence (HCE-AI) in smart factories within the framework of Industry 5.0. Industry 5.0 emphasizes collaboration between humans and advanced technologies to create sustainable, people-centered production environments. The vast amount of data generated by factory sensors presents opportunities for machine learning models to enhance predictive maintenance. However, challenges such as latency, cybersecurity risks, and the lack of transparency in AI decision-making persist. This chapter highlights research perspectives like Explainable AI (XAI), Federated Learning, Self-supervised Learning, 1-bit machine, and advanced quantization techniques to address these issues. With its ability to process data locally, Edge AI reduces latency and enhances real-time response capabilities. This article also discusses the role of Digital Twins in optimizing performance and predictive maintenance. By leveraging these technologies, smart factories can achieve higher efficiency, safety, and sustainability, thus aligning with the human-centric in the 5.0 revolution.

Keywords Explainable artificial intelligence · Self-supervised learning · Edge artificial intelligence · Federated learning · Digital twin · Smart wearable · Human-centered edge AI

1 Introduction

The Industrial Revolution 4.0 has profoundly transformed smart factories through the widespread adoption of technologies such as the Internet of Things (IoT), Artificial Intelligence (AI), and Big Data. These advancements have led to automation systems,

L. H. Nguyen · X. Zeng · K. P. Tran
ULR 2461 - GEMTEX - Génie et Matériaux Textiles, ENSAIT, University of Lille, 59000 Lille, France

L. H. Nguyen · K. D. Tran (✉) · K. P. Tran
International Chair in DS & XAI, International Research Institute for Artificial Intelligence and Data Science, Dong A University, Danang, Vietnam
e-mail: ductk@donga.edu.vn

K. P. Tran (ed.), *Artificial Intelligence for Safety and Reliability Engineering*, Springer Series in Reliability Engineering,
https://doi.org/10.1007/978-3-031-71495-5_5

industrial robots, and connectivity solutions that create a manufacturing environment that is more efficient, flexible, and intelligent than ever before. However, progress does not stop there. The world is now transitioning into the 5.0 Industrial Revolution, marking a new step forward with a higher vision. This new era focuses on the cooperation between people and technology to create a sustainable and people-centered production environment.

Industry 5.0 emphasizes the interaction between humans and machines, where technology supports and enhances human creativity and innovation [1]. Smart factories in the 5.0 era aim to optimize performance and minimize costs while fostering a friendly and safe working environment, thereby improving the quality of life for workers. The integration of artificial intelligence and the Internet of Things enables factories' automation and digitalization processes to become more flexible, minimize errors, and increase control and transparency in the production process [1, 2].

The rest of this chapter is structured as follows: We introduce artificial intelligence that can participate in the monitoring and control levels of the factory system in part 2, thereby presenting advanced algorithms. advances in machine learning to transparently explain the decision-making rationale of artificial intelligence, we present a self-supervised learning approach for predictive maintenance when indirect data is limited but still delivering precision in abnormal predictive maintenance of equipment, we then refer to Edge Artificial Intelligence combined with Federated Learning methodology to ensure real-time response in predictive maintenance and ensure uplift High data security capabilities, We refer to Digital Twin to enable simulation and decision support in predictive maintenance. Part 3 addresses the difficulties and challenges and presents perspectives for applying human-centered Edge AI in smart factories. We propose acceleration solutions for software and hardware. Use FPGA when implementing the Federated Learning method. We propose solutions that integrate AI algorithms into Digital Twin to secure cyber attacks on factory control systems. And gave the opinion that smart wearable devices integrate AI to protect the health and safety of workers during work. In Sect. 4, we aim to develop a robust framework for detecting multivariate time series anomalies in a coupled environment to enhance Edge AI applications. This chapter ends with "Concluding Remarks" in Sect. 5, summarizing the main recommendations in Smart Factory 5.0.

2 State of the Art

Recent advancements in technologies such as IoT Gateway, 5G wireless networks, Artificial Intelligence, and smart embedded systems have significantly impacted various fields, including agriculture [3], healthcare [4], and industrial plants [5]. The development of smart sensor devices has increased the volume of data sent to the cloud, where AI models can analyze this data for demand prediction and production optimization. Analyzing and predicting potential failures and risks to prevent system shutdowns is crucial in industrial plants. Home automation systems are typically divided into five levels: Field level, Control level, Monitoring level, Management level, and Enterprise level [6]. AI can support and enhance operations from the control level upwards to ensure effective system performance. The levels to be considered are as follows:

- **Control level:** AI can support automated control systems by employing machine learning algorithms to optimize control strategies and detect and respond to abnormalities or issues [7].
- **Supervisor level:** At this level, AI can analyze data from various sources to provide an overview of the system's state. It can identify trends, predict potential problems, and recommend remedies before issues occur [8].
- **Management and Enterprise level:** AI can aid in decision-making and strategic planning. By analyzing big data and providing detailed analytical reports, AI helps management make data-driven decisions, optimize business processes, and improve overall organizational efficiency [9] (Fig. 1).

Another critical aspect is the transition from traditional isolated control systems to open systems (Internet-based) for data collection, monitoring, and remote control, which enhances flexibility and ease of use. However, this transition also introduces threats from network vulnerabilities, affecting system control. Therefore, applying AI for control and monitoring is essential. AI can analyze and process collected data to intervene and warn about potential attacks on the system, predicting risks that could cause sudden stops and reduce unwanted damage to the plant. Consequently, AI has become indispensable in automation systems within smart factories; see, [1, 10–12], among many others.

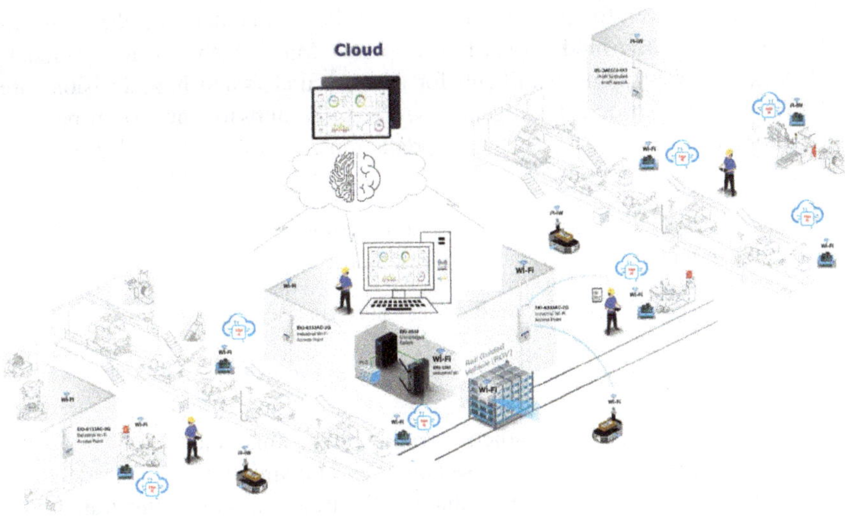

Fig. 1 The architecture of a smart factory

By integrating AI into control levels in automation systems, AI uses predictive analytics to foresee equipment failures and maintenance needs, allowing for proactive maintenance planning. This enhances control and transparency in the production process.

With the advancement of AI in deep learning, algorithms such as Recurrent Neural Networks (RNN) [13, 14], Convolutional Neural Networks (CNN) [15], Transformers [16], Long-Short Term Memory (LSTM) [17, 18], and Generative Adversarial Networks (GAN) [19] have been applied in smart factories to increase accuracy in predictive maintenance. These AI models use data from on-device sensors to predict abnormalities and potential risks to the system. Additionally, algorithms can predict the remaining usage time of equipment during operation. The combination of Federated Learning [20] and smart embedded devices [21] enhances analysis capabilities, real-time feedback, and energy savings during computation. These promising results will soon be applied in real smart factories [22]. The following sections will delve into the concepts and techniques used in this chapter.

2.1 *Explainable Artificial Intelligence*

Predictive maintenance of industrial equipment's health status is widely applied through data obtained from sensors [2, 23]. Machine learning models can detect and diagnose errors automatically, issuing warnings to prevent risks and reduce system downtime. Advanced machine learning models like CNNs, Transformers, and Reinforcement Learning models [24, 25] offer significant potential for smart factory applications. However, for users to trust AI in predictive maintenance, the decision-making process of these models must be transparent. Many machine learning models are "black boxes", making it difficult for users to understand how decisions are made. Explainable Artificial Intelligence (XAI) [26, 27] aims to provide transparency and detailed explanations for AI model predictions. XAI can be divided into two categories:

- Transparent models: These models, such as Linear Regression and Decision Trees, have built-in features that explain their decisions.
  - Linear Regression provides predictions based on a learned function:

$$\mathbf{y} = \mathbf{w}^1 x^1 + \mathbf{w}^2 x^2 + \mathbf{w}^3 x^3 + \cdots + \mathbf{w}^n x^n + b \tag{1}$$

    Users can understand the predictions based on this function.
  - Decision Trees: Decisions are based on rules, allowing users to understand the model's prediction results by examining the tree's structure after training on data [28]. Therefore, the inherent properties of these models provide explanations for their decisions. However, these models generally have lower prediction performance than advanced models such as artificial neural networks.

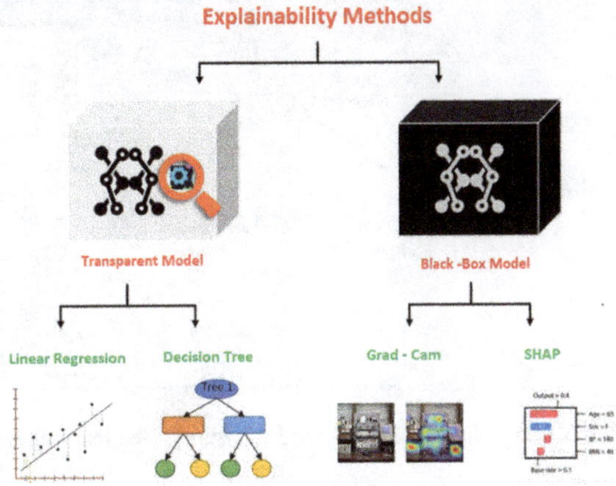

Fig. 2 Transparent and Black-Box models with explainable AI

- Black-box models with post-hoc explainability: These techniques develop separate models to explain the predictions of the original model. Methods include Visualization Explanation with Grad-CAM [29] and Feature Relevance with Shapley Additive exPlanations (SHAP) [30] (Fig. 2).

XAI enhances the transparency and interpretability of AI models, ensuring confidence in their outputs. Various XAI strategies facilitate model interpretation, such as SHAP, Local Interpretable Model-agnostic Explanations (LIME) [31] and Grad-CAM. After training the deep learning, a model, the XAI module, can be integrated to gain better insights and improve our understanding of the model's behavior. However, we also need efforts to develop Explainable-by-Design AI techniques that can overcome the disadvantages of post-host techniques in many cases [32].

2.2 *Self-supervised Learning*

Self-supervised learning (SSL) [33] has gained attention due to its ability to learn from unlabeled data, reducing the need for extensive labeled datasets. This approach addresses challenges associated with data annotation, making SSL relevant in fields such as computer vision, natural language processing, and anomaly detection [34–36].

SSL has shown promise in anomaly detection and predictive maintenance. By leveraging unlabeled data, SSL models learn representations and capture underlying patterns, enhancing anomaly detection and improving generalization to unseen instances. For example, in [37], a Transformer-based SSL framework was used for pavement anomaly detection, improving performance without explicitly labeled data and enabling real-time edge detection. In [38], a framework for unsupervised anomaly

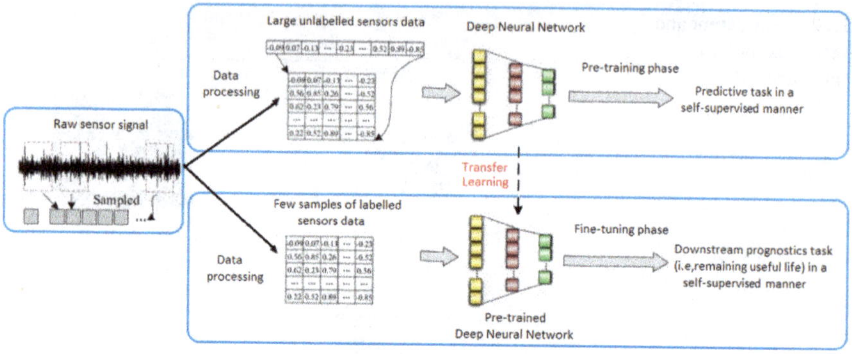

Fig. 3 Flow chart of the proposed self-supervised learning framework

detection in images utilized SSL and dynamic local augmentation, achieving competitive performance. In [39], a deep adversarial anomaly detection method combined SSL and deep adversarial training, improving performance over strong baselines.

Predictive maintenance is crucial for extending equipment life and preventing unscheduled downtime, which can lead to significant losses. For instance, automotive manufacturers face high costs due to unscheduled downtime [40]. Smart sensors, industrial IoT, and AI in predictive maintenance help collect and analyze data accurately and continuously, providing anomalies and transparent explanations to build user trust [41].

Numerous studies applying AI with sensor data, such as CNN [42] and LSTM [43] can estimate abnormalities before failures occur. Remaining Useful Life Prediction (RUL) [44, 45] predicts the remaining useful life of devices, optimizing maintenance planning and minimizing downtime. SSL models [46] learn from unlabeled data, reducing the reliance on labeled datasets and making SSL favored in various fields. SSL models for RUL prediction include two stages (Fig. 3):

- **Pre-training:** The prediction model is trained on raw unlabelled data to learn features.
- **Fine-tuning:** The pre-trained model is trained on labeled data in a supervised manner.

These models outperform non-pre-trained models in RUL prediction tasks, showing promise for predictive maintenance with limited labeled data [46].

2.3 *Edge Artificial Intelligence*

Edge Artificial Intelligence (Edge AI) involves deploying artificial intelligence directly on devices or at the "edge" of the network rather than transmitting data to a central data center for processing. This approach enables devices and systems to process data and deploy machine learning models where the data is generated without relying on a network connection to a central server. In smart factories, data generated by sensors during production line operations is typically connected and transmitted to the cloud for AI processing to solve specific problems, such as detecting operational abnormalities or product defects. This cloud-based approach, however, can introduce delays. Real-time responses can be provided by applying AI models directly on Edge AI devices.

The advancement of integrated circuits, smart embedded systems, and reprogrammable circuits allows users to customize and reconfigure their functions. Field Programmable Gate Arrays (FPGAs), with their grid of logic blocks, can perform multiple tasks simultaneously and be configured for specific computing requirements [47]. FPGAs as Edge AI devices can execute computational and decision-making tasks previously only possible on powerful computing systems. By configuring AI within the processor chip, FPGAs can run AI algorithms directly at the network edge, meeting real-time requirements and significantly reducing the bandwidth needed for data streaming to the cloud. Additionally, developing machine learning models optimized for Edge AI reduces the difficulty of deployment, enhancing efficiency when implementing these devices in smart factories (Fig. 4).

Federated learning is essential to enhance Edge AI devices in factories. Federated Learning is a distributed machine learning method where model algorithms are trained on multiple devices or servers without moving raw data to a central data center. Instead, local devices perform computations and send only model updates (such as model parameters) to a central server, where they are aggregated to improve the global model. This approach enhances data privacy and security, minimizes data transfer bandwidth, and enables efficient AI operations on Edge AI devices with limited computational capabilities, see [12, 20, 48]. Federated learning risks data poisoning attacks, where inaccurate or malicious data is intentionally added to local

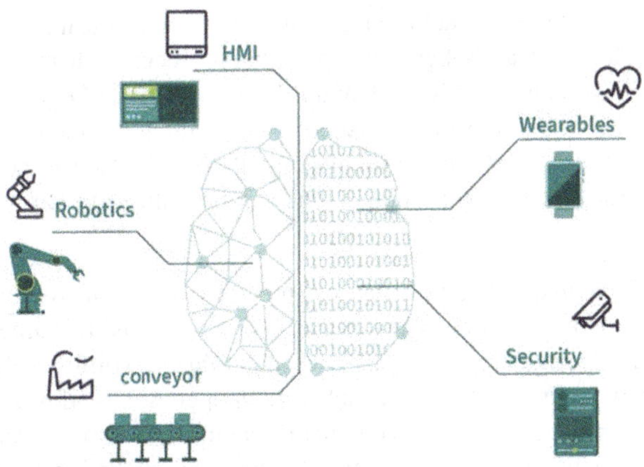

Fig. 4 Federated learning-based edge AI model

devices to undermine or deceive the global model. This risk can lead to factories or organizations being cautious about using federated learning due to concerns about security and reliability. However, advanced models have been developed to mitigate the risk of such data poisoning attacks. Security measures such as data integrity checks, anomaly detection, and secure machine learning techniques are employed to protect federated learning models from these attacks. As a result, Edge AI has become more feasible and secure, allowing factories to leverage the benefits of federated learning without worrying about security threats [49–51].

Another approach to improve computing speed and train AI models on Edge AI devices with limited memory resources is quantization. This method involves rounding or truncating values, such as converting 0.8–1 or 0.2–0. However, quantization can lead to a loss of accuracy and may require retraining the network or using higher bit-precision quantization. A recent approach using minimum mean square error (MMSE) for both weights and activations allows low bit-precision inference without retraining the entire network. This method supports 4-bit, 2-bit, and 1-bit integer quantization, making deploying pre-trained models on limited hardware resources feasible. This approach significantly reduces memory and computational requirements, optimizing hardware for Edge AI, see [52–55], among many others.

Edge AI devices are placed close to data generation points, enabling easy user interaction, allowing human control over AI decisions, and fostering a sense of safety and autonomy. Therefore, AI algorithms on Edge AI must operate on principles that explain their decisions, ensuring users can understand and trust them. This approach not only focuses on developing smart systems but also places people at the center of the production process in the factory, see [1, 56–59], among many others.

2.4 *Digital Twin*

A Digital Twin is a digital model of an actual system created using sensor data and other sources. This field holds significant potential for recent industrial applications, particularly when combined with IoT and AI. Specifically, a Digital Twin creates an accurate digitized version of a physical system, allowing manufacturers, engineers, or maintenance personnel to monitor, predict, and optimize performance in real-time [60, 61]. Digital Twin applications can be utilized in various industries, including:

- Predictive Maintenance: Digital Twins can predict and prevent problems or failures in physical systems, optimizing maintenance processes and resource utilization.
- Simulation and Design: Digital Twins can simulate and test different designs before building a new system, saving time and costs in operations and production.
- Optimize Performance: Digital Twins enable manufacturers to optimize real-world system performance by continuously collecting and analyzing data from smart sensors, minimizing waste, and enhancing productivity.

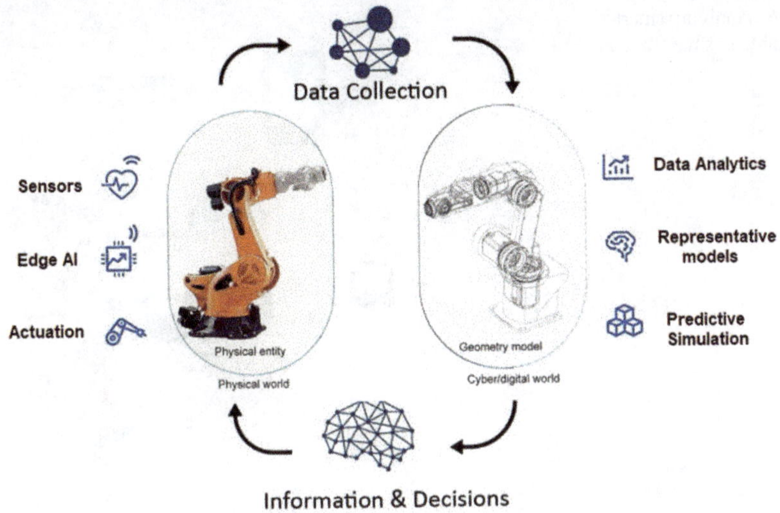

Fig. 5 Digital twin in smart factory

- Decision Support: Digital Twins provide real-time information on system operations, aiding managers and decision-makers in making smart, data-driven decisions (Fig. 5).

2.5 *Smart Wearable*

In factories, engineers and technical staff are responsible for supervising and ensuring the safe operation of production lines. Their health and safety need to be protected at all times, and this is where wearable technology comes into play. Wearable devices, worn on the body or attached to clothing, are designed to collect data and monitor health and activity on the factory floor. These devices include smartwatches, which can connect to mobile phones and offer features such as notifications, activity tracking, heart rate monitoring, and health support applications. Fitness trackers monitor physical activity, including steps taken, distance traveled, and energy expended, helping operators monitor their health while working. Equipping factories with these wearable devices offers several benefits. Workers feel protected and secure, which enhances their ability to contribute effectively and increases their long-term commitment to the company. For business owners, the data gathered and analyzed from these devices can optimize work processes, reduce operating costs, and boost production efficiency (Fig. 6).

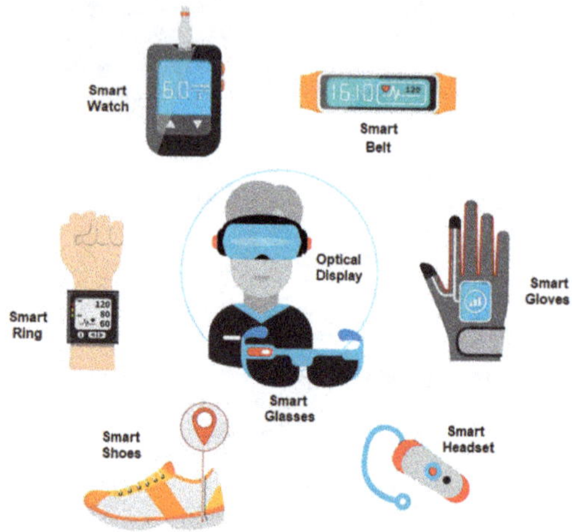

Fig. 6 Applying smart wearable in smart factory

3 Difficulties, Challenges, and Perspectives for Application of Human-Centered Edge Artificial Intelligence for Smart Factory Applications in Industry 5.0

The challenges and limitations hindering artificial intelligence (AI) use in smart factories today are as follows: AI research often relies on black-box models, making the decision-making process opaque and difficult to apply. Predictive maintenance requires labeled data; however, fault data for equipment is very scarce. Furthermore, concentrating data on the cloud/server to run AI models poses cybersecurity threats and fails to meet real-time requirements for the system. Most algorithms have not yet been optimized to accelerate AI through hardware design. In addition, the automation system in the factory is programmed and controlled without human interaction when there is a problem that requires decision-making in the factory has not been done care. Based on the above limitations, we proposed designing a framework for deploying AI applications in smart factories. Integrate Explainable AI to make model decisions transparent and combine self-supervised learning and XAI technique for predictive maintenance with limited labeled data. Using Edge AI devices (FPGA) combined with 1-bit ML method to lightweight data but still have high accuracy and can be deployed on Federated learning. Integrate AI into Digital Twin to predict and simulate potential cybersecurity threats.

3.1 Human-Centered Edge AI

To meet the demands of Industry 5.0, we suggest the following innovative solutions for resource optimization and energy conservation in smart factories:

- Edge AI: Edge AI reduces the pressure on network data transmission and central systems while ensuring factory data security. Edge AI can detect product defects, perform predictive maintenance, and support decision-making and health monitoring for factory employees. This approach minimizes latency and enhances real-time response capabilities.
- Federated Learning: Explainable By Design Edge AI approaches in federated learning to enhance model accuracy without centralizing data can be developed, thus preserving privacy and reducing the computational load on central servers. This method ensures efficient and secure data processing [62–64].
- Data Quantization: Data quantization techniques, such as reducing data to 1 or 2-bit representations, can be implemented to minimize energy consumption and speed up computations on Edge AI devices. This method allows efficient processing on devices with limited computational resources without compromising accuracy [65, 66].
- Analog Deep Learning: Analog Deep Learning can be developed for smart factory applications. Analog deep learning can achieve high computational speeds with minimal energy consumption, making it a promising approach for handling large datasets in smart factories [67, 68].

3.2 FPGA Design for the Acceleration of Edge Artificial Intelligence

Deep learning models deployed to edge devices handle significant data, such as image processing for defect detection and vibration analysis for motor failure prediction. The limited resources on edge devices present challenges in efficiently executing these models. We propose using Field-Programmable Gate Arrays (FPGAs) to accelerate deep learning tasks on edge devices:

- Parallel Processing: The parallel processing capabilities of FPGAs to accelerate deep neural network (DNN) tasks effectively can be used to develop AI approaches. This capability is critical for real-time applications in smart factories.
- Energy Efficiency: FPGAs consume less power than conventional processors (CPUs, GPUs), making them suitable for low-power edge applications [69].
- Optimization Techniques: The optimization of hardware/software structure on FPGAs can be implemented using low-bit representations like binary to reduce hardware resource usage, enhancing processing speed and energy efficiency. Converting high-level algorithm descriptions to hardware description languages (HDL) can further optimize FPGA implementations [70].

- Data Quantization: Data quantization approximation algorithms can be deployed to FPGAs to ensure efficient execution of AI tasks on edge devices with limited computational resources [71, 72].

3.3 *Predictive Maintenance with Human-Centered Edge AI*

Predictive maintenance is critical for preventing unplanned downtime and ensuring the safety and efficiency of smart factories. The continuous monitoring of equipment and environmental conditions can be done by strategically placing Edge AI devices throughout the factory. These devices can analyze sensor data to predict and prevent potential failures, ensuring smooth and uninterrupted operations Herrmann and Pfeiffer [73].

- Self-supervised Learning (SSL): Self-supervised learning techniques can be proposed to predict the remaining useful life of equipment in maintenance. With this technique, measured data from raw sensors will still give accurate prediction results when labeled data is scarce in industrial factories.
- Explainable AI (XAI): Integration of XAI into machine learning systems to ensure transparent decision-making processes, enabling engineers and staff to understand and verify AI-generated recommendations can be deployed. This transparency is crucial for maintaining trust and ensuring the reliability of AI systems in critical operations.
- Edge devices: Edge devices for the Federated Learning approach can be implemented to limit the amount of data sent to the center, ensure real-time response, and save system energy.

By following these steps, factories can effectively implement predictive maintenance, leveraging Edge AI devices to ensure safe and reliable operations.

3.4 *PLC Cybersecurity and Fault Detection with Edge AI-Driven Digital Twin*

Protecting Programmable Logic Controller (PLC) and Supervisor Controller and Data Acquisition (SCADA) systems from cyber attacks is essential for maintaining the integrity of smart factories. Enhancing cybersecurity using AI to detect anomalies and prevent potential threats can be proposed. This proactive approach ensures production processes' continuous and safe operation, minimizing the risk of costly disruptions. Building a Digital Twin system integrating AI algorithms allows simulating scenarios and testing preventive measures in a virtual environment before applying them in practice, thereby allowing the detection of potential risks or signs of attacks in real systems [61, 74].

- AI Integration: Incorporating AI algorithms in the digital twin [75] to predict and identify potential cyber threats can be implemented. Establish a warning mechanism for detected abnormalities in the SCADA system and network infrastructure.
- PLC has AI function: Integrating an embedded system with AI functions on the PLC controller can be proposed so that the PLC can perform the Logic control function and be used as an AI edge device. We can perform many measures using artificial intelligence right on the PLC, such as predictive maintenance of equipment failures, prevention of network attacks, etc. [76].

In this way, we can effectively improve the security of PLC, SCADA, and network security systems in smart factories using Digital Twin technology controlled by Edge AI. This approach ensures strong protection against cyber threats and minimizes the risk of system disruption.

3.5 Human-Centered Edge AI and Wearable Technology for Workplace Health and Safety

Wearable devices with Edge AI can monitor employees' health and safety in real time. These devices can provide feedback on environmental conditions, working posture, and fatigue levels, helping to prevent accidents and ensure a safe working environment. We will integrate multiple features into a wearable device to enhance comfort and usability while prioritizing data privacy and security [77, 78].

- We use AI-integrated edge devices to make smart wearables [79], these devices are trained with AI models to analyze health data to detect abnormalities, unsafe movements, or hazards from the surrounding environment to ensure worker safety during work.
- We will propose a wearable device that is compact and convenient for users but still meets the real-time accuracy needed when implementing AI algorithms on the device (Fig. 7).

To increase speed on wearable devices and data privacy, we will apply a federated learning approach to ensure wearables are widely used in smart factories soon.

In addition, smart wearable devices need to be integrated with Explainable AI algorithms to increase the transparency of the AI model, which will bring reliability to users. This is considered a human-centered factor in the 5.0 industrial revolution.

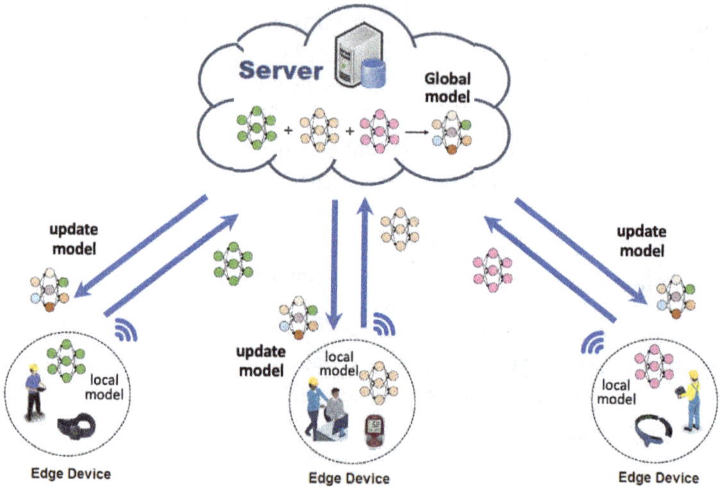

Fig. 7 Applying smart wearable in smart factory

4 Future Direction

Our future work aims to develop a robust framework for multivariate time series anomaly detection in a federated setting, specifically for intelligent manufacturing. Due to the efficiency of transformer-based models recently [48, 80, 81], we propose to utilize these advanced models in combination with traditional classification algorithms OCSVM [82]/SVDD [83], Support Vector Machine (SVM) [84] to detect anomalies effectively. The transformer-based models will extract critical information, leveraging their ability to handle long-range dependencies and parallel processing. The classification algorithms will then identify abnormalities based on these features. Federated learning will ensure data privacy by training models locally on edge devices and aggregating them on a central server without transferring raw data. To enhance interpretability, we propose integrating XAI techniques to understand better the deep learning models used for anomaly detection. We will also explore self-supervised learning techniques to address the challenge of limited labeled data, using the data's inherent structure to learn meaningful representations. This approach improves the performance of anomaly detection models and enhances their interpretability. Additionally, we will incorporate 1-bit machine learning to optimize the models, reducing computational and memory requirements for efficient deployment in resource-constrained environments.

4.1 *Transformer-Based Anomaly Detection in Smart Manufacturing*

First, to prepare the data for analysis, we will normalize the multivariate time series data and segment it into fixed-size windows. This preprocessing ensures that the input to the models is consistent, allowing the transformer-based frameworks to learn and extract critical information directly from the raw data.

- **Model architectures:** For the model architectures, we propose several transformer-based approaches, such as combining Transformer architecture with Variational Autoencoders (VAEs) [85], or the combination of transformer with Temporal Convolutional Networks (TCNs) [86], and select the most suitable one for our application.
  - **Transformer-based VAEs:** will employ transformer layers in both the encoder and decoder to handle long-range dependencies and parallel processing. The encoder outputs a distribution (mean and variance) for each input, mapping the data to a probabilistic latent space, typically Gaussian. The decoder then reconstructs the input data from the sampled latent distribution. Variational inference is used to learn the data distribution, enabling the generation of new samples and capturing the underlying structure.
  - **Transformer-based TCN:** integrates transformer layers with TCN layers to efficiently handle long-term dependencies and temporal patterns. Temporal convolutions are utilized to capture temporal dependencies, and the sequence is reconstructed from the encoded representation using both transformer and TCN layers.

  We will evaluate these architectures to determine the most effective approach for our specific application.
- **Classification Algorithms:** For detecting the abnormal objects from the inputs, we suggest using OCSVM, SVDD, or SVM, which is expected to reduce the computation time and improve the accuracy in detecting abnormal data. This classifier will be integrated with the selected transformer-based model for anomaly detection.
- **Self-supervised learning approach:** SSL has shown promise in anomaly detection by leveraging unlabeled data to learn representations and capture underlying patterns, enhancing the model's ability to detect anomalies and generalize to unseen instances. Our objective in upcoming work is to develop a self-supervised learning [46] framework for detecting anomalies in multivariate time series data within a federated setting, with applications in intelligent manufacturing. We leverage SSL techniques to enhance anomaly detection and enable efficient and accurate detection, even with limited or inaccessible labeled data across multiple devices.
- **1-bit machine learning:** we will also incorporate 1-bit machine learning (1-bit ML) [66] into our robust transformer-based model for detecting anomalies in multivariate time series data within a federated setting. This approach, with applications

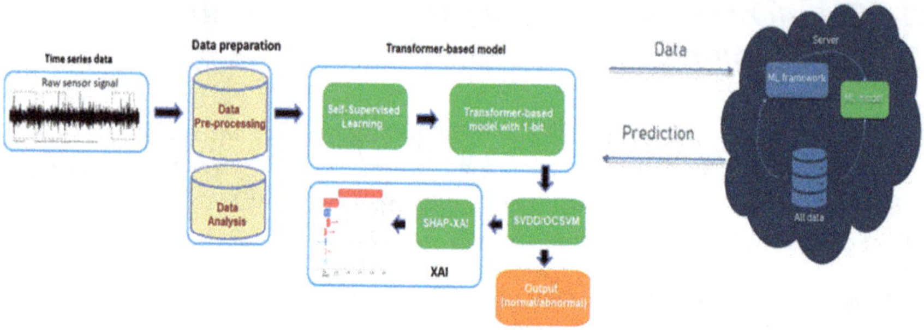

Fig. 8 Proposed robust transformer-based model

in intelligent manufacturing, significantly reduces model size and computational requirements by quantizing weights and activations to a single bit. By leveraging 1-bit ML, we aim to optimize the efficiency and performance of our models. This is particularly beneficial in scenarios where data is collected through sensors and processed on edge devices with limited computational power and memory. Integrating 1-bit ML ensures that our anomaly detection models remain lightweight and suitable for deployment in resource-constrained environments without compromising accuracy.

- **Explainable Artificial Intelligence (XAI):** Explainable Artificial Intelligence (XAI) is crucial for improving the transparency and interpretability of AI models. By using XAI algorithms, we can better understand how these models operate, increasing confidence in their predictions. XAI allows us to elucidate the workings of the AI model, its intended effects, and any potential biases it might have. Various XAI techniques, such as SHAP [30] and Local Interpretable Model-agnostic Explanations (LIME) [31], have been introduced in the literature to aid in model interpretation. In this proposal, following the training of the transformer-based model, we plan to integrate one of these XAI methods to gain better insights and enhance our understanding of the model's behavior (Fig. 8).

4.2 *FPGA-Based Acceleration for Edge AI in Smart Manufacturing*

The integration of FPGAs in edge AI applications offers significant performance and energy efficiency benefits. Our future work focuses on leveraging FPGAs to enhance deep learning models for real-time applications in smart manufacturing. We will pursue the following advancements:

- **Dynamic Reconfiguration:** We propose exploring dynamic reconfiguration techniques to allow FPGAs to adapt their architecture based on task requirements, improving resource efficiency [87, 88].
- **Adaptive Quantization:** We will develop adaptive quantization techniques to balance accuracy and efficiency in real-time processing, ensuring optimal performance for different tasks [89, 90].

These advancements aim to develop robust and efficient FPGA-based edge AI systems, enhancing their use in intelligent manufacturing.

5 Concluding Remarks

Summarizing the above, we present our research on advanced machine learning methods such as self-supervised learning, Explainable AI, and Federated Learning. These methods offer practical solutions to the problem of scarcity of labeled data and the risks of data loss in predictive maintenance at industrial plants. We stress the importance of explainability in machine learning models, which empowers humans in system operation. Furthermore, we propose practical solutions to accelerate Edge AI devices, enhancing flexibility and real-time response for the system. The key components are as follows:

- **Integrating Explainable AI and Self-Supervised Learning:** To enhance transparency and accuracy in anomaly detection and Remaining Useful Life (RUL) predictions, we will integrate XAI techniques to provide insights into model decisions and use self-supervised learning to leverage inherent data structures for better representation learning. This combined approach will improve model performance and interpretability.
- **Incorporating 1-Bit Machine Learning:** To optimize models for deployment in resource-constrained environments, we will incorporate 1-bit machine learning techniques, reducing computational and memory requirements and making the models more efficient for edge deployment.
- **Deploying Federated Learning:** We will enable efficient and secure distributed machine learning by deploying federated learning. This approach trains models locally on edge devices and aggregates them on a central server without transferring raw data. It will reduce energy consumption and preserve data privacy.
- **Utilizing FPGA-Based Edge AI Devices:** To achieve real-time, low-energy computation, we will deploy FPGA-based edge AI devices [91], enhancing the performance of AI applications in smart factories by providing efficient, localized processing.
- **Implementing Digital Twin Technology:** We will implement Digital Twin technology to create virtual simulations of physical systems, allowing operators to observe, simulate, and proactively address potential issues, thus enabling proactive maintenance and operational efficiency.

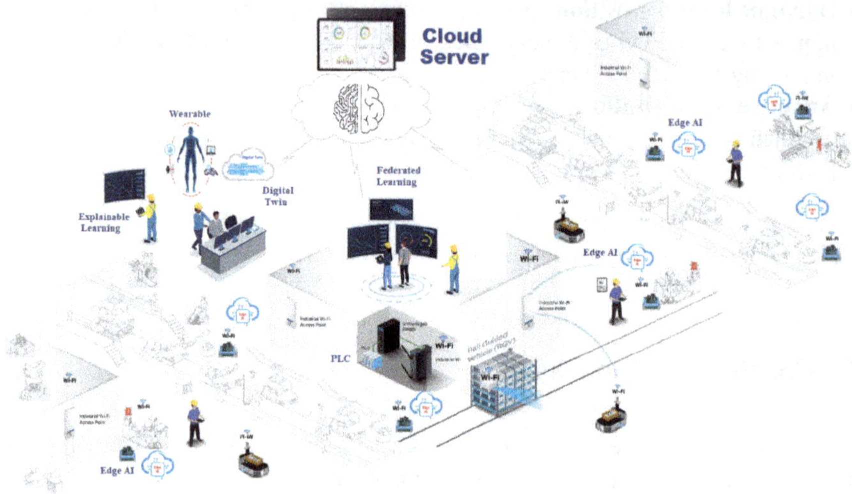

Fig. 9 Model smart factory in industrial 5.0

– **Developing Advanced Smart Wearable Devices:** For comprehensive health and safety monitoring, we will develop advanced smart wearable devices that provide real-time data on workers' health and safety conditions, ensuring a proactive approach to workplace safety (Fig. 9).

By integrating these advanced techniques and technologies, our framework aims to address key challenges in intelligent manufacturing, enhancing anomaly detection, data privacy, and resource efficiency. This comprehensive approach integrates advanced machine learning and AI techniques to enhance predictive maintenance, improve system efficiency, and ensure data privacy.

To make artificial intelligence a reality in smart factories with the goal of human-centered beings, we propose to develop interfaces and interactive tools to seamlessly integrate AI with the system, enabling easy monitoring of AI operations in smart factory 5.0. This comprehensive approach aims to advance AI applications in smart factories, promote a human-centered Industry 5.0 revolution, and foster a safer, more efficient, and sustainable industrial environment.

References

1. Nguyen DH, Tran KP (2023) Artificial intelligence for smart manufacturing in industry 5.0: Methods, applications, and challenges. In: Artificial intelligence for smart manufacturing: methods, applications, and challenges, pp 5–33
2. Nguyen DH, Tran KP, Zeng X, Koehl L, Castagliola P, Bruniaux P (2019) Industrial internet of things, big data, and artificial intelligence in the smart factory: a survey and perspective. In: ISSAT international conference on data science in business, finance and industry, pp 72–76

3. Khayyam H, Javadi B, Jalili M, Jazar RN (2020) Artificial intelligence and internet of things for autonomous vehicles. In: Nonlinear approaches in engineering applications: automotive applications of engineering problems pp 39–68
4. Alavi AH, Jiao P, Buttlar WG, Lajnef N (2018) Internet of things-enabled smart cities: state-of-the-art and future trends. Measurement 129:589–606
5. Shariatzadeh N, Lundholm T, Lindberg L, Sivard G (2016) Integration of digital factory with smart factory based on internet of things. Procedia Cirp 50:512–517
6. Dey C, Sen SK (2020) Industrial automation technologies. CRC Press
7. Ma R, Kareem SW, Kalra A, Doewes RI, Kumar P, Miah S (2022) Optimization of electric automation control model based on artificial intelligence algorithm. Wireless Commun & Mobile Comput
8. Udo W, Muhammad Y (2021) Data-driven predictive maintenance of wind turbine based on scada data. IEEE Access 9:162370–162388
9. Kaggwa S, Eleogu TF, Okonkwo F, Farayola OA, Uwaoma PU, Akinoso A (2024) Ai in decision making: transforming business strategies. Int J Res Sci Innov 10(12):423–444
10. Shah Y, Sengupta S (2020) A survey on classification of cyber-attacks on iot and iiot devices. In: 2020 11th IEEE annual ubiquitous computing, electronics & mobile communication conference (uemcon). IEEE
11. Rani B, Palmer GM, Kathrine GJW, Ewards SEV, et al (2022) Intrusion detection system for cyber attacks in food and beverage industry. In: 2022 international conference on automation, computing and renewable systems (ICACRS), pp 1287–1291. IEEE
12. Truong TH, Ta PB, Dao ML, Tran DL, Nguyen MD, Bui DT, Tran KP et al (2021) Detecting cyberattacks using anomaly detection in industrial control systems a federated learning approach. Comput Ind 132:103509
13. Chu Y, Fei J, Hou S (2019) Adaptive global sliding-mode control for dynamic systems using double hidden layer recurrent neural network structure. IEEE Trans Neural Netw Learn Syst 31(4):1297–1309
14. Markiewicz M, Wielgosz M, Bocheński M, Tabaczyński W, Konieczny T, Kowalczyk L (2019) Predictive maintenance of induction motors using ultra-low power wireless sensors and compressed recurrent neural networks. IEEE Access 7:178891–178902
15. Li Z, Liu F, Yang W, Peng S, Zhou J (2021) A survey of convolutional neural networks: analysis, applications, and prospects. IEEE Trans Neural Netw Learn Syst 33(12):6999–7019
16. Han K, Xiao A, Wu E, Guo J, Xu C, Wang Y (2021) Transformer in transformer. Adv Neural Inf Proc Syst 34:15908–15919
17. Graves A, Graves AL (2012) Long short-term memory. In: Supervised sequence labelling with recurrent neural networks, pp 37–45
18. Essien A, Giannetti C (2020) A deep learning model for smart manufacturing using convolutional lstm neural network autoencoders. IEEE Trans Ind Inf 16(9):6069–6078
19. Kusiak A (2020) Convolutional and generative adversarial neural networks in manufacturing. Int J Prod Res 58(5):1594–1604
20. Li L, Fan Y, Tse M, Lin KY (2020) A review of applications in federated learning. Comput & Ind Eng 149:106854
21. Seng KP, Lee PJ, Ang LM (2021) Embedded intelligence on fpga: survey, applications and challenges. Electronics 10(8):895
22. Nguyen DH, Tran PH, Do TH, Tran KP (2023) Quality control for smart manufacturing in industry 5.0. In: Artificial intelligence for smart manufacturing: methods, applications, and challenges, pp 35–64. Springer
23. Zonta T, Costa CAD, Righi RDR, Lima MJD, Trindade ESD, Li GP (2020) Predictive maintenance in the industry 4.0: a systematic literature review. Comput & Ind Eng 150:106889
24. Rodríguez MLR, Kubler S, Giorgio AD, Cordy M, Robert J, Traon YL (2022) Multi-agent deep reinforcement learning based predictive maintenance on parallel machines. Robot Comput-Integr Manuf 78:102406
25. He Z, Tran KP, Thomassey S, Zeng X, Xu J, Yi C (2021) A deep reinforcement learning based multi-criteria decision support system for optimizing textile chemical process. Comput Ind 125:103373

26. Ta PB, Do TH, Tran KD, Tran KP (2023) Explainable articial intelligence for cybersecurity in smart manufacturing. In: Artificial intelligence for smart manufacturing: methods, applications, and challenges. Springer, pp 199–223
27. Do TH, Ta PB, Tran KD, Tran KP (2023) Efficient and trustworthy federated learning-based explainable anomaly detection: challenges, methods, and future directions. In: Artificial intelligence for smart manufacturing: methods, applications, and challenges. Springer, pp 145–166
28. Mahbooba B, Timilsina M, Sahal R, Serrano M (2021) Explainable artificial intelligence (xai) to enhance trust management in intrusion detection systems using decision tree model. Complexity 1–11:2021
29. Selvaraju RR, Cogswell M, Das A, Vedantam R, Parikh D, Batra D (2017) Grad-cam: visual explanations from deep networks via gradient-based localization. In: Proceedings of the IEEE international conference on computer vision, pp 618–626
30. Lundberg SM, Lee SI (2017) A unified approach to interpreting model predictions. In: Advances in neural information processing systems, vol 30. Curran Associates, Inc
31. Ribeiro MT, Singh S, Guestrin C (2016) why should i trust you?: explaining the predictions of any classifier. In: Proceedings of the 22nd ACM SIGKDD international conference on knowledge discovery and data mining, pp 1135–1144
32. Das A, Rad P (2020) Opportunities and challenges in explainable artificial intelligence (xai): a survey. arXiv:2006.11371
33. Jaiswal A, Babu AR, Zadeh MZ, Banerjee D, Makedon F (2020) A survey on contrastive self-supervised learning. Technologies 9(1):2
34. Ramesh S, Srivastav V, Alapatt D, Yu T, Murali A, Sestini L, Nwoye CI, Hamoud I, Sharma S, Fleurentin A et al (2023) Dissecting self-supervised learning methods for surgical computer vision. Med Image Anal 102844
35. Baevski A, Hsu W-N, Xu Q, Babu A, Gu J, Auli M (2022) Data2vec: A general framework for self-supervised learning in speech, vision and language. In: International conference on machine learning. PMLR, pp 1298–1312
36. Park D, Ahn CW (2019) Self-supervised contextual data augmentation for natural language processing. Symmetry 11(11):1393
37. Lin Z, Wang H, Li S (2022) Pavement anomaly detection based on transformer and self-supervised learning. Autom Constr 143:104544
38. Yoa S, Lee S, Kim C, Kim HJ (2021) Self-supervised learning for anomaly detection with dynamic local augmentation. IEEE Access 9:147201–147211
39. Zhang X, Mu J, Zhang X, Liu H, Zong L, Li Y (2022) Deep anomaly detection with self-supervised learning and adversarial training. Pattern Recogn 121:108234
40. Resende C, Folgado D, Oliveira J, Franco B, Moreira W, Oliveira AJ, Cavaleiro A, Carvalho R (2021) Tip4.0: industrial internet of things platform for predictive maintenance. Sensors 21(14):4676
41. Pang G, Aggarwal C (2021) Toward explainable deep anomaly detection. In: Proceedings of the 27th ACM SIGKDD conference on knowledge discovery & data mining, pp 4056–4057
42. Murphy J (2016) An overview of convolutional neural network architectures for deep learning. Microway Inc, pp 1–22
43. Fischer T, Krauss C (2018) Deep learning with long short-term memory networks for financial market predictions. Eur J Oper Res 270(2):654–669
44. Sikorska JZ, Hodkiewicz M, Ma L (2011) Prognostic modelling options for remaining useful life estimation by industry. Mech Syst Signal Proc 25(5):1803–1836
45. Ferreira C, Gon G (2022) Remaining useful life prediction and challenges: a literature review on the use of machine learning methods. J Manuf Syst 63:550–562
46. Akrim A, Gogu C, Vingerhoeds R, Sala M (2023) Self-supervised learning for data scarcity in a fatigue damage prognostic problem. Eng Appl Artif Intell 120:105837
47. Vermesan O, Coppola M (2023) Embedded edge intelligent processing for end-to-end predictive maintenance in industrial applications. In: Industrial artificial intelligence technologies and applications. River Publishers, pp 157–175

48. Raza A, Tran KP, Koehl L, Li S (2023) Anofed: adaptive anomaly detection for digital health using transformer-based federated learning and support vector data description. Eng Appl Artif Intell 121:106051
49. Nuding F, Mayer R (2020) Poisoning attacks in federated learning: an evaluation on traffic sign classification. In: Proceedings of the tenth ACM conference on data and application security and privacy, pp 168–170
50. Raza A, Li S, Tran KP, Koehl L (2022) Using anomaly detection to detect poisoning attacks in federated learning applications. arXiv:2207.08486
51. Nguyen TD, Rieger P, Miettinen M, Sadeghi AR (2020) Poisoning attacks on federated learning-based iot intrusion detection system. In: Proceedings of the workshop decentralized IoT systems and security (DISS), vol 79
52. Nayak P, Zhang D, Chai S (2019) Bit efficient quantization for deep neural networks. In: 2019 fifth workshop on energy efficient machine learning and cognitive computing-NeurIPS edition (EMC2-NIPS). IEEE, pp 52–56
53. Zhou J, Wu J, Gao Y, Ding Y, Tao C, Li B, Tu F, Cheng KT, So HKH, Wong N (2023) Dybit: dynamic bit-precision numbers for efficient quantized neural network inference. IEEE Trans Comput-Aided Des Integr Circuits Syst 123:123
54. Wang P, Hu Q, Zhang Y, Zhang C, Liu Y, Cheng J (2018) Two-step quantization for low-bit neural networks. In: Proceedings of the IEEE conference on computer vision and pattern recognition, pp 4376–4384
55. Ma S, Wang H, Ma L, Wang L, Wang W, Huang S, Dong L, Wang R, Xue J, Wei F (2024) The era of 1-bit llms: all large language models are in 1.58 bits. arXiv:2402.17764
56. Shneiderman B (2022) Human-centered AI. Oxford University Press
57. Riedl MO (2019) Human-centered artificial intelligence and machine learning. Hum Behav Emerg Technol 1(1):33–36
58. Schmidt A (2020) Interactive human centered artificial intelligence: a definition and research challenges. In: Proceedings of the international conference on advanced visual interfaces, pp 1–4
59. Xu W (2019) Toward human-centered ai: a perspective from human-computer interaction. Interactions 26(4):42–46
60. Huang H, Yang L, Wang Y, Xu X, Lu Y (2021) Digital twin-driven online anomaly detection for an automation system based on edge intelligence. J Manuf Syst 59:138–150
61. Aivaliotis P, Georgoulias K, Chryssolouris G (2019) The use of digital twin for predictive maintenance in manufacturing. Int J Comput Integr Manuf 32(11):1067–1080
62. Nguyen DH, Nguyen HL, Kieu NH, Nguyen VH, Truong TH, Tran KP (2023) Trans-lighter: a light-weight federated learning-based architecture for remaining useful lifetime prediction. Comput Ind 148:103888
63. Soares EA (2022) Explainable-by-design deep learning. Lancaster University (UK)
64. Angelov P (2021) Keynote: explainable-by-design deep learning. In: 2021 IEEE international conference on pervasive computing and communications workshops and other affiliated events (PerCom Workshops)
65. Pham P, Abraham JA, Chung J (2021) Training multi-bit quantized and binarized networks with a learnable symmetric quantizer. IEEE Access 9:47194–47203
66. Khobahi S, Soltanalian M (2020) Model-based deep learning for one-bit compressive sensing. IEEE Trans Signal Proc 68:5292–5307
67. Tsai H, Ambrogio S, Narayanan P, Shelby RM, Burr GW (2018) Recent progress in analog memory-based accelerators for deep learning. J Phys D: Appl Phy 51(28):283001
68. Onen M, Emond N, Wang B, Zhang D, Ross FM, Li J, Yildiz B, Alamo JAD (2022) Nanosecond protonic programmable resistors for analog deep learning. Science 377(6605):539–543
69. Belabed T, Coutinho MGF, Fernandes MA, Sakuyama CV, Souani C (2021) User driven fpga-based design automated framework of deep neural networks for low-power low-cost edge computing. IEEE Access 9:89162–89180

70. Haris J, Gibson P, Cano J, Agostini NB, Kaeli D (2021) Secda: efficient hardware/software co-design of fpga-based dnn accelerators for edge inference. In: 2021 IEEE 33rd international symposium on computer architecture and high performance computing (SBAC-PAD). IEEE, pp 33–43
71. Reddy KM, Vasantha MH, Kumar YN, Gopal CK, Dwivedi D (2021) Quantization aware approximate multiplier and hardware accelerator for edge computing of deep learning applications. Integration 81:268–279
72. Nain G, Nain P, Sharma KK, Sharma GK (2022) Towards edge computing in intelligent manufacturing: past, present and future. J Manuf Syst 62:588–611
73. Herrmann T, Pfeiffer S (2023) Keeping the organization in the loop: a socio-technical extension of human-centered artificial intelligence. AI & Soc 38(4):1523–1542
74. Kandasamy NK, Venugopalan S, Wong TK, Leu NJ (2022) An electric power digital twin for cyber security testing, research and education. Comput Electr Eng 101:108061
75. Huang Z, Shen Y, Li J, Fey M, Brecher C (2021) A survey on ai-driven digital twins in industry 4.0: smart manufacturing and advanced robotics. Sensors 21(19):6340
76. Milik A (2016) On hardware synthesis and implementation of plc programs in fpgas. Microprocess Microsyst 44:2–16
77. Patel V, Chesmore A, Legner CM, Pandey S (2022) Trends in workplace wearable technologies and connected-worker solutions for next-generation occupational safety, health, and productivity. Adv Intell Syst 4(1):2100099
78. Nguyen T, Tran KD, Raza A, Nguyen QT, Bui HM, Tran KP (2023) Wearable technology for smart manufacturing in industry 5.0. In: Artificial intelligence for smart manufacturing: methods, applications, and challenges. Springer, pp 225–254
79. Kalapothas S, Flamis G, Kitsos P (2022) Efficient edge-ai application deployment for fpgas. Information 13(6):279
80. Vaswani A, Shazeer N, Parmar N, Uszkoreit J, Jones L, Gomez AN, Kaiser L, Polosukhin I (2017) Attention is all you need. In: Advances in neural information processing systems, vol 30. Curran Associates, Inc
81. Nguyen TTV, Heuchenne C, Tran KD, Tran KP (2023) A novel transformer-based anomaly detection approach for ecg monitoring healthcare system. In: International conference on safety and security in IoT. Springer, pp 111–129
82. Schölkopf B, Williamson RC, Smola A, Taylor JS, Platt J (1999) Support vector method for novelty detection. In: Advances in neural information processing systems, 12
83. Tax DM, Duin RP (2004) Support vector data description. Mach Learn 54:45–66
84. Pisner DA, Schnyer DM (2020) Support vector machine. In: Machine learning. Elsevier, pp 101–121
85. Kingma DP, Welling M, et al (2019) An introduction to variational autoencoders. Found Trends® Mach Learn 12(4):307–392
86. Lea C, Vidal R, Reiter A, Hager GD (2016) Temporal convolutional networks: a unified approach to action segmentation. In: Computer vision–ECCV 2016 workshops: Amsterdam, The Netherlands, October 8–10 and 15–16, 2016, Proceedings, Part III 14. Springer, pp 47–54
87. Pezzarossa L, Schoeberl M, Sparsø J (2017) A controller for dynamic partial reconfiguration in fpga-based real-time systems. In: 2017 IEEE 20th international symposium on real-time distributed computing (ISORC), pp 92–100. IEEE
88. Pezzarossa L, Kristensen AT, Schoeberl M, Sparsø J (2018) Using dynamic partial reconfiguration of fpgas in real-time systems. Microprocess Microsyst 61:198–206
89. Jiao L, Luo C, Cao W, Zhou X, Wang L (2017) Accelerating low bit-width convolutional neural networks with embedded fpga. In: 2017 27th international conference on field programmable logic and applications (FPL). IEEE, pp 1–4
90. Baskin C, Liss N, Zheltonozhskii E, Bronstein AM, Mendelson A (2018) Streaming architecture for large-scale quantized neural networks on an fpga-based dataflow platform. In: 2018 IEEE international parallel and distributed processing symposium workshops (IPDPSW). IEEE, pp 162–169
91. Xu C, Jiang S, Luo G, Sun G, An N, Huang G, Liu X (2020) The case for fpga-based edge computing. IEEE Trans Mobile Comput 21(7):2610–2619

Enhancing Trustworthiness in AI-Based Prognostics: A Comprehensive Review of Explainable AI for PHM

Duc An Nguyen, Khanh T. P. Nguyen, and Kamal Medjaher

Abstract Prognostics and Health Management (PHM) has emerged as an essential field for ensuring the reliability and safety of critical systems, increasingly becoming a key component in their maintenance and operation. While artificial intelligence (AI) has become a powerful tool for PHM, the opaque nature of AI methods might present hurdles in their deployment for real-world PHM applications, as the lack of transparency in how these methods process and interpret data. In this light, Explainable AI (XAI) is a rapidly growing field that aims to address those challenges. This understanding enhances the trustworthiness and acceptability of AI-based PHM solutions. Nevertheless, inconsistency in terminologies and concepts across diverse studies on XAI in PHM persists, accompanied by the absence of a standardized taxonomy. Our study aims to address this gap by presenting a comprehensive review of XAI in PHM, clarifying key concepts, and introducing a structured taxonomy tailored for PHM applications. This taxonomy facilitates the effective categorization of XAI methods, guiding their selection based on specific PHM requirements. We also delve into the delicate balance between the model's explainability and performance, emphasizing XAI's role in ensuring algorithmic fairness and reliability in PHM solutions. Additionally, the paper outlines practical challenges and potential future research directions in XAI for PHM. Overall, this contribution significantly advances the understanding and application of XAI in PHM practices.

Keywords Explainable AI · Machine learning · Deep learning · Prognostics and health management · Litterature review · Condition monitoring

---

D. A. Nguyen (✉) · K. T. P. Nguyen · K. Medjaher
The National Engineering School of Tarbes - INP Toulouse, Production Engineering Laboratory, 65000 Tarbes, France
e-mail: duc_an.nguyen@enit.fr

K. T. P. Nguyen
e-mail: thi-phuong-khanh.nguyen@enit.fr

K. Medjaher
e-mail: kamal.medjaher@enit.fr

K. P. Tran (ed.), *Artificial Intelligence for Safety and Reliability Engineering*, Springer Series in Reliability Engineering,
https://doi.org/10.1007/978-3-031-71495-5_6

1 Introduction

Prognostics and Health Management (PHM) has emerged in engineering as a promising scientific discipline and has become an indispensable component in product life cycles [1–3]. It harnesses condition monitoring data, complemented by insights from operational practitioners, to detect anomalies, diagnose faults, and predict the Remaining Useful Life (RUL) of systems [2]. Such proactive measures facilitate maintenance scheduling, which is crucial for maintaining system health, thereby ensuring system availability, reliability, and safety [1]. Fundamental tasks within PHM involve the development of health indicators, diagnostics, fault detection, and prognostics. These tasks are crucial for detecting potential issues early, accurately diagnosing system conditions, and forecasting system longevity.

The PHM field encompasses various applications, from simple components like rolling bearings and hard disk drives to complex systems, including steam generators, chemical processes, and turbine engines. This extensive scope highlights the increasing interest and commitment to advancing PHM, as evidenced by the growing volume of research activities, illustrated in Fig. 1. This figure indicates a discernible trend of exponential growth in Artificial Intelligence (AI) enabled PHM methods in recent years [2]. These AI-based PHM methods are renowned for their proficiency in learning from extensive datasets to make precise predictions, gradually supplanting traditional model-based approaches [4]. This shift is conspicuous in the increasing

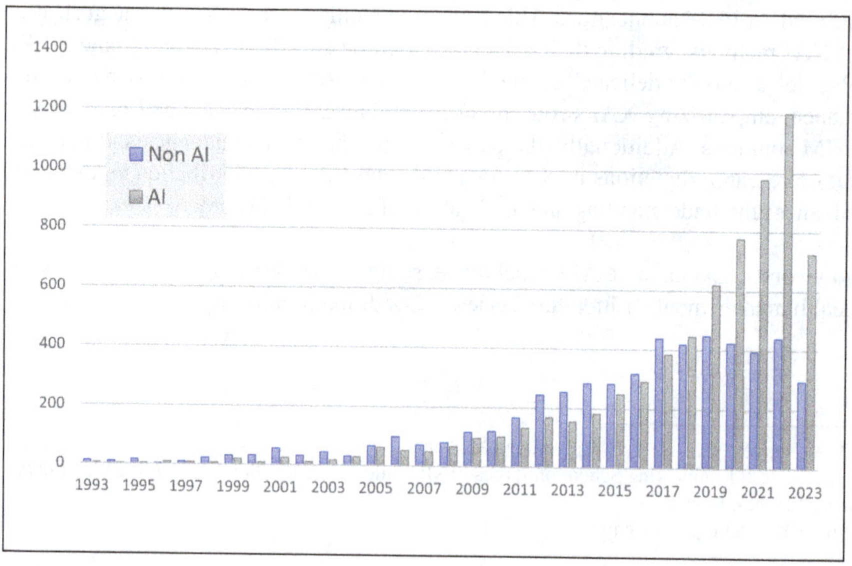

Fig. 1 Development trend of AI and non-AI based PHM studies (until 8/2023) [2]

preference for AI methods in PHM, owing to their superior ability to handle complex, nonlinear, and high-dimensional data, characteristics frequently encountered in real-world systems.

Nonetheless, the integration of AI into PHM does not come without challenges. One significant drawback of AI-based PHM methods is their lack of interpretability. The "black-box" nature of many AI algorithms makes it difficult for users and stakeholders to understand and trust the decision-making process [5, 6]. XAI has emerged as a pivotal research area, addressing the need to clarify AI outputs for enhanced human comprehension [7]. It strives to explain the internal decision-making mechanisms of AI, striking a balance between model explainability and performance efficiency. Hence, AI systems can articulate their reasoning and operational logic from data alone, enhancing transparency without needing supplementary prior knowledge [8, 9]. In addition, XAI is instrumental in ensuring algorithmic fairness, identifying biases, and guaranteeing the intended performance of AI models, thereby upholding algorithmic accountability [10]. The insights afforded by XAI into AI's operational mechanisms enable users to trust and effectively manage AI-driven systems, especially in the context of PHM, where precise and reliable predictions are vital for maintenance and operational planning.

Despite its advantages, XAI in PHM is still an emerging field with various approaches and taxonomies, each with strengths and limitations. Many of these approaches, however, suffer from inconsistencies and lack a unified framework specifically tailored for PHM applications. In this chapter, we aim to bridge this gap by providing a comprehensive review of XAI within the context of PHM. We will clarify key concepts and terminologies in XAI and present a taxonomy specifically designed for its application in PHM. *The main goal is to* ***demystify XAI for PHM practitioners*** *and* ***provide a clear path forward for integrating these advanced AI techniques into PHM practices****, ensuring not only improved performance but also enhanced trust and understanding.*

The rest of this chapter is organized into the following sections. In Sect. 2, we delineate the literature review methodology and provide an overview of relevant prior research. This is the foundation for our study, asserting the significant contributions to the field. Section 3 delves into the fundamentals of XAI, clarifying key terminologies and examining the relationship between accuracy and interpretability. This section also outlines the general taxonomies and frameworks of XAI. Proceeding further, Sect. 4 introduces a specific taxonomy of XAI tailored for PHM, discussing various concepts, categorizing methods, and evaluating the pros and cons of each approach. Section 5 analyzes XAI trends in PHM, focusing on state-of-the-art methods commonly employed for each PHM task. In Sect. 6, we discuss the challenges faced and future directions in this area. Section 7 concludes the chapter by summarizing the significant contributions of this study and discussing its perspectives.

2 Litterature Review Methodology and Related Review Works

This section outlines the methodology employed in our literature research, accompanied by a comprehensive survey of related works in the field of XAI in PHM. Our focus is critically analyzing existing literature. Through this examination, we aim to highlight both the strengths and limitations of current approaches, demonstrating how these insights have shaped and motivated our research. Specifically, we seek to identify gaps in the existing body of knowledge, setting the stage for our contributions to advance the understanding and application of XAI in PHM.

2.1 *Methodology of Literature Review*

Figure 2 outlines the methodology of our literature review, comprising three primary stages: problem definition and search, screening, and detailed information extraction from selected papers. We utilized the Web of Science (WoS) platform for its reliability and extensive database as the primary research tool for document retrieval.

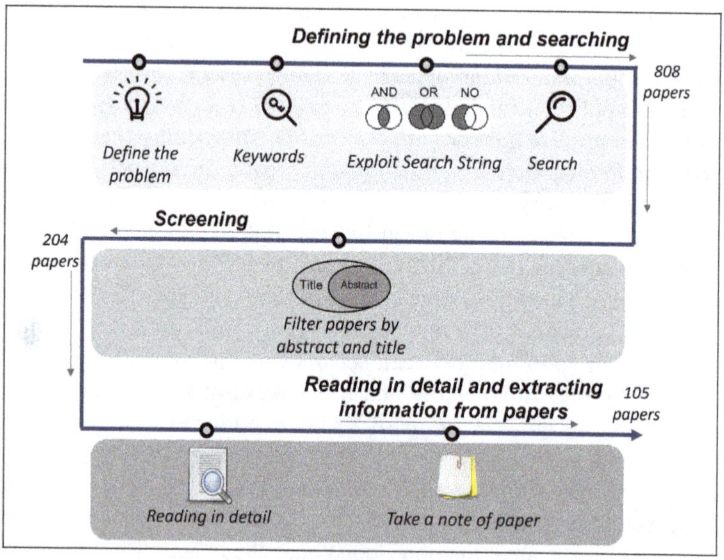

Fig. 2 The stages of the conducted literature review

The literature review process encompasses the following stages in detail:

1. ***Defining the problem and searching:*** at this stage, we identified our research focus, seeking keywords pertinent to data-driven PHM methods (including the PHM tasks like health indicators, fault detection, diagnostics, and prognostics) integrated with XAI. These keywords, spanning three fields: PHM, AI, and XAI, were synthesized to formulate our search query. To ensure a concentration on contemporary methods, we restricted our search to publications from 2013 to August 2023. Utilizing Web of Science's (WoS) advanced search options, we refined our search by excluding irrelevant keywords, categories, and non-credible publishers. This approach yielded 808 papers, encompassing both relevant and extraneous studies.
2. ***Screening:*** this step involves filtering papers based on their titles, keywords, and abstracts. Initially, we skim each paper to check for unrelated keywords indicative of other fields or to determine if it is focused on datasets or patents. Papers with content irrelevant to our research focus will be excluded. This is essential to ensure our review is confined to documents within the PHM field. Next, we determine whether the papers pertain to data-driven or model-based methods in the context of PHM. This distinction is often challenging, as not all papers explicitly state their methodology in the abstract, keywords, or title. We then examine the papers for indications of XAI integration, focusing on those mentioning XAI-related keywords. Nevertheless, abstracts and titles may not always provide sufficient information for definitive categorization, necessitating further detailed examination in subsequent steps. At this juncture, the selection of papers relevant to our review reduced the number to 204.
3. ***Extracting the detailed information:*** the initial screening, based on information extracted from titles, abstracts, and keywords, proved insufficient for comprehensively identifying all relevant XAI studies in PHM. Notably, AI-enabled papers can use model-based methods to provide feature interpretability, which may not be explicitly mentioned in their titles or abstracts. Hence, a thorough review of these papers at this stage is crucial. Additionally, we eliminated studies from unrelated fields that failed to demonstrate their pertinence to our research. Throughout this phase, we continuously updated our Excel database, recording key details such as paper titles, authors, specific PHM tasks addressed, AI and XAI methodologies employed, data sources, and a summary of the central themes. This rigorous selection process resulted in the identification of 105 papers for in-depth analysis in the subsequent stages of our review.

2.2 Related Literature Review Works

Integrating XAI into PHM has garnered significant attention, aiming to enhance the transparency and trustworthiness of predictive maintenance models. Recognizing the need for a consolidated understanding in this rapidly evolving field, comprehensive literature reviews have played a crucial role. They distill individual studies into coherent narratives, identify prevailing trends, and benchmark the current state of the art. This subsection discusses the insights and frameworks these reviews have contributed to the field and clarifies how they pave the way for our research.

A notable example is the work of Sharma et al. [11], authors provide a review on condition-based maintenance using machine learning, emphasizing the role of interpretability, particularly in anomaly detection within PHM. Their review introduces a sub-taxonomy based on model-agnostic methods, underscoring the emergence of XAI as a crucial field in addressing interpretability challenges in complex models. However, it is limited in scope, focusing only on anomaly detection tasks and lacking a comprehensive taxonomy for broader PHM applications. In the realm of building energy management, Chen et al. [12] present a state-of-the-art review, categorizing studies into ante-hoc and post-hoc interpretability approaches. Their work emphasizes the importance of interpretable machine learning in improving model understanding and trust. However, this paper's focus on specific case studies building energy limits its applicability in the broader PHM domain, pointing to the need for more inclusive research. Carter et al. [13] conduct a systematic review in process industries, exploring the use of interpretable machine learning and the challenges posed by the opaque nature of these models in high-risk scenarios. The review stresses the necessity of interpretability for safety and trust but primarily concentrates on surrogate models for feature importance, offering limited insight into broader XAI concepts. In another study, the review on anomaly explanation by Yepmo et al. [14] provides a fresh perspective by covering XAI to explain anomaly detection tasks. However, its scope extends beyond PHM, exploring anomaly detection and explanation in various fields, thereby highlighting the necessity for more focused research within PHM. A comprehensive review from Cheng et al. [15] about predictive maintenance in the context of Industry 4.0 discusses the role of visual analytics's role in the PdM framework, including XAI. While it underscores the use of visual analytics and XAI in enhancing machine learning models for anomaly detection, it does not provide detailed taxonomies or classifications specific to XAI methods, indicating an area for further thorough exploration. Additionally, a systematic review of feature relevance in anomaly detection through XAI approaches offers a broad categorization of various methods is also presented by Tritscher et al. [16]. It provides a comprehensive view that covers data-specific, perturbation-based, gradient-based, and model-specific explanations. Nevertheless, its general approach across multiple domains, not explicitly tailored to PHM, suggests a research gap in focusing on PHM-specific methodologies.

The above reviews substantially advance the comprehension of XAI in PHM, yet *they expose research deficiencies, particularly in establishing detailed taxonomies and investigating domain-specific applications*. Although these studies cover a range of XAI categories, there are evident shortcomings. Certain categories are **either insufficient or redundant, and some fail to meet the distinct demands of PHM tasks**. Additionally, much of the current research either **overly focuses on specific PHM aspects or treats PHM as a minor element within a larger field of study**. A notable issue is the **inconsistency and ambiguity in XAI terminology, which**

complicates the development of precise taxonomies for PHM and impedes a unified understanding of the field. By addressing these gaps, our work presents an exhaustive examination of XAI, offering multiple perspectives. We aim therefore to formulate a clear, detailed taxonomy tailored to PHM, thus rectifying the limitations identified in prior literature and proposing a more systematic and pertinent framework for XAI's application in PHM scenarios.

3 Understanding XAI Fundamentals

This section aims to elucidate key terminological distinctions in XAI and provide an overview of XAI taxonomies. Section 3.1 distinguishes between explainability and interpretability in XAI, also addressing related terminologies. Section 3.2 discusses the relationship between interpretability and accuracy in XAI. Finally, Sect. 3.3 details the classification methodologies of XAI methods from various perspectives.

3.1 XAI Terminologies

XAI is a research field focused on making the results of AI systems more understandable to humans. It encompasses approaches designed to provide insights into the decision-making processes of AI, exploring the reasoning behind these decisions, highlighting the system's strengths and limitations, and offering predictions about its future behavior [7, 8]. XAI is particularly significant in addressing AI's transparency and trust issues, as it attempts to balance model interpretability with efficiency [7, 8]. A fundamental characteristic of XAI is its capacity to produce explanations directly from data and the possibility to use external information or prior knowledge to enhance and validate the reliability of its outcomes. This approach, known as data-driven XAI, emphasizes the importance of data-centric methods in deriving explanations [9]. However, it is essential to note that there is no single, universally agreed-upon definition of XAI. It is often perceived more as a movement aimed at improving AI transparency and trust rather than a strictly defined technical term [7]. The objective of XAI research extends beyond creating AI models capable of providing explanations for their outputs; it also seeks to deepen the comprehension of these explanations, moving past basic definitions and purposes. This endeavor includes examining fundamental concepts that define the XAI field and understanding the diverse perspectives on explainability across various research communities [7, 17]. Therefore, for those researching in XAI, acquiring a thorough understanding of these core concepts and the broader significance of explainability in AI is essential [7].

3.1.1 Explainability and Intepretability

In the XAI field, certain terminologies often lead to confusion. A notable example is the distinction, or lack thereof, between interpretability and explainability. Many studies treat these concepts as synonymous, using them interchangeably [7].

Miller [18] views interpretability and explainability as criteria for evaluating how comprehensible AI decisions are to humans in specific contexts. Adadi et al. [7] explore XAI's applications across various fields, such as transportation and healthcare, suggesting that explainability is a subset of interpretability. Similarly, Carletti et al. [19] use "interpretability" to cover both concepts. This interchangeable usage is also noted by Amarasinghe et al. [20] and Sharma et al. [11].

However, some research, like that of Gergely Hajgató et al. [21], distinguishes the two, suggesting that interpretability focuses on cause-and-effect relationships, while explainability delves into the algorithm's internal mechanisms. Ferraro et al. [22] offer another perspective, defining interpretability as the capacity to make AI understandable to humans and explainability as the interface facilitating this understanding.

In our analysis, the definition of XAI concepts varies depending on the application or perspective, presenting a challenge in distinguishing these terms. Within the PHM field's confines, refining these concepts is crucial, as each term offers different insights into XAI, leading to varied explanations of the model and aiding users in accurately understanding XAI's functionality. Inspired by the study presented in [17], we clarify the differences between fundamental XAI terms, including explainability and interpretability in Fig. 3. The main differences between explainability and interpretability are as follows:

- *Interpretability:* this refers to a machine learning model's ability to offer insights into its behavior and structure, aiming to shed light on internal processes during prediction. Interpretability can arise from the model-building phase or data analysis, serving as a passive attribute indicating a model's clarity to human observers [23]. Essentially, it relates to models with transparent decision-making processes [24].
- *Explainability:* defined as a model's ability to elucidate its operations to users [23, 25], explainability is a more encompassing concept than interpretability. If a model is not inherently interpretable, methods are employed to demystify its decision logic, providing additional information to simplify its workings for the audience [24, 26]. Explainability includes a model's innate interpretability and external methods enhancing its understandability. While an interpretable model can be considered inherently explainable, the reverse is not necessarily true. Explainability can be categorized into scientific and human-centric. Scientific explainability aids developers in improving models, identifying biases, and building trust, enhancing the model's quality. Human-centric explainability focuses on user trust and understanding, ensuring that model processes are communicated effectively.

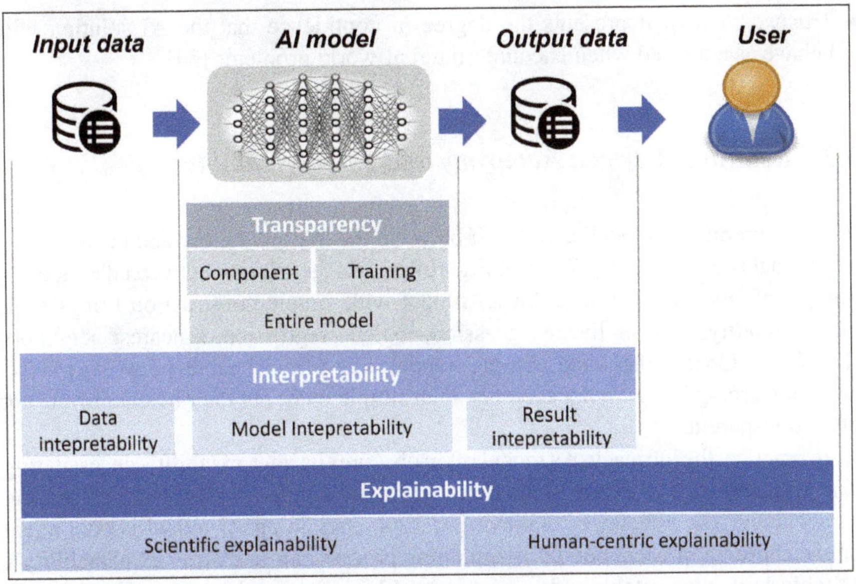

Fig. 3 Difference between fundamental terms in XAI

3.1.2 Other Related Terms

In the context of XAI, apart from explainability and interpretability, terms like transparency, comprehensibility, understandability, and trustworthiness are frequently employed to describe diverse aspects of the technology. However, the nuanced differences and possible overlap among these concepts necessitate a focused approach. Thus, our study will center on terminologies that are not only widely recognized but also pertinent in the XAI domain, particularly for PHM applications. The key terms under consideration are as follows:

- *Transparency*: this term is employed to characterize a model's potential comprehensibility and its contrast with the concept of a "black-box" model [7, 17, 23, 27]. In essence, it quantifies the degree of accessibility to either the data or the model itself [8]. Transparency can be categorized into three distinct levels, namely: model transparency (pertaining to the entire model), design transparency (concerning model components), and algorithmic transparency (relating to the training algorithms) [17, 23, 28].
- *Comprehensibility*: it refers to the ability of a learning algorithm to articulate its acquired knowledge in a manner comprehensible to humans [23, 29–32].
- *Understandability*: it demonstrates how a model can facilitate an individual's comprehension of its functioning and processes without necessitating an explanation of its internal structure or the precise algorithms employed for internal data processing [23, 33].

- *Trustworthiness*: it presents the degree of confidence that the AI solution will behave as expected when encountering real-world problems [34].

3.2 Relation Between Accuracy and Interpretability

For each prediction model, a trade-off between the interpretability and the accuracy of a model is essential [12, 23, 35]. Figure 4 shows the relation between the model's interpretability and accuracy. An AI model with simple construction brings high interpretability, such as linear regression, logistic regression, k-nearest neighbors (KNN), etc. On the other hand, complex models like artificial neural networks (ANN) or deep learning bring higher accuracy than simple ones, but their mechanism is not often transparent.

In practice, the intricacy of a model often obscures its interpretability, necessitating various techniques to expound upon the AI model's functionality while preserving its accuracy. The efficacy of achieving transparency in an AI model is contingent on the chosen approach, but this overarching process can be called "explainability". Notably, both interpretable and non-interpretable models can be rendered explainable through specific methods of explanation. However, it is important to emphasize that "interpretability" cannot be retroactively imposed upon a non-interpretable model.

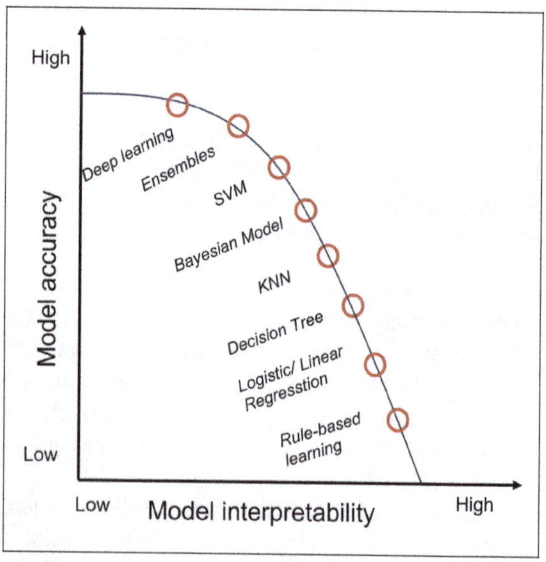

Fig. 4 Relation between the interpretability and accuracy in an AI model

3.3 *Comprehensive Taxonomy of XAI Methods from Different Perspectives*

XAI approaches can be classified into distinct groupings contingent upon various methodological perspectives. As depicted in Fig. 5, a comprehensive delineation of these taxonomies underscores the multifaceted nature inherent in XAI approaches. Specifically, this exposition will delve into six perspectives that underlie the categorization of XAI methodologies: (1) the stage at which AI is employed, (2) the degree of dependency on the AI model, (3) the scope of explainability, (4) the internal XAI mechanisms, (5) the task executed by the AI model, and (6) the nature of the data input to the AI model. These delineated categories not only serve as a structured framework for classification but also exert a substantial influence on the implementation and evolutionary trajectory of XAI techniques. They offer a roadmap for developers and researchers, facilitating the reasonable selection, deployment, and enhancement of XAI methodologies that align with specific contextual or operational requirements. Furthermore, it is worth noting that an XAI methodology may concurrently belong to multiple categories contingent upon the particular vantage points of the observer. Our study covers a holistic perspective, thereby affording flexibility in developing and applying XAI across diverse research paradigms.

1. ***AI stage:*** XAI approaches can be split into distinct groups based on the stages at which XAI is integrated into the AI model's lifecycle. This encompasses three key phases: (1) preceding the model's construction, denoted as "pre-model methods", (2) concurrent with the model's development, referred to as "during-model methods", and (3) after the model's construction, encompassing "post-hoc methods" or "post-model methods".

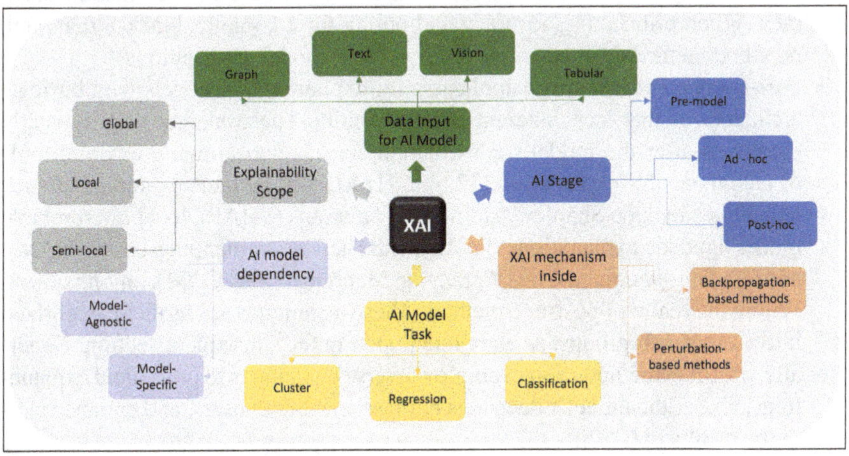

Fig. 5 Comprehensive XAI taxonomy from different perspectives

- *Pre-model methods*: this refers to the process of examining, comprehending, and delineating data through the utilization of explainable data analysis, comprehensive dataset documentation, succinct dataset summarization, transparent feature engineering techniques, unbiased data collection, and meticulous experimental design. These practices aim to establish clarity in the data before the model's construction, as highlighted by Joshi [36]. Pre-model approaches are usually autonomous and do not depend upon a specific model or system architecture to apply it [37]. Common examples of these approaches are principal component analysis and t-distributed stochastic neighbor embedding [37, 38]. This work aids users in comprehending the data, enabling the formulation of strategies to construct or fine-tune AI models for greater appropriateness.
- *Ad-hoc methods*: also known as ante-hoc methods, including intrinsic interpretable or transparent models. These models are inherently self-explanatory and typically feature interpretable architectures [12, 39–41]. They can be classified into two subgroups, partially interpretable and completely interpretable models, discussed in Sect. 4.2. This approach integrates transparency into machine learning or deep learning models during construction or training. Let us take linear regression as an example: it is a straightforward model used to predict continuous outcomes based on input variables. It is therefore easy to understand and explain because it works by linearly combining input variables [12, 42]. Within intricate deep learning models, interpretability is attained through the utilization of the latent space to produce saliency maps, elucidating attention mechanisms, and applying signal processing methodologies [43–49]. These techniques are employed to augment the partial interpretability of such models. In summary, ad-hoc approaches improve users' understanding of the model's structure and results, thereby strengthening the credibility of the model or algorithm [41, 50]. However, it's crucial to acknowledge that although models in this category provide interpretability, there might be a compromise in accuracy. When prioritizing accuracy and opting for a complex black-box model, post-hoc methods can be employed to furnish model explanations.
- *Post-hoc methods*: they are applicable for explaining AI models with intricate architectures that lack inherent interpretability. These methods are typically employed after the model's construction stage, often utilizing external tools or techniques for explanation [12, 40, 41, 51, 52]. Notable examples of such techniques include Shapley Additive exPlanations (SHAP), local interpretable model-agnostic explanations (LIME), class activation mapping-based (CAM), and Gradient-weighted Class Activation Mapping (Grad-CAM), among others. Within the realm of post-hoc methods, these techniques are further categorized into various sub-groups, as elaborated upon in the subsequent section. Generally, post-hoc methods aim to employ diverse techniques to offer lucid explanations, render the model's decisions comprehensible to users, and enhance trust in the results [41, 50].

2. ***AI model dependency:*** in this particular classification methodology, which hinges upon the degree of reliance of XAI methods on the AI model's architecture,

the XAI methods can be systematically classified into two subcategories. These categories encompass model-agnostic methods, which possess the versatility to be employed with any machine learning model, and model-specific methods, which are expressly tailored for specific model configurations.

- *Model-Agnostic group*: it covers XAI methods that apply to all types of machine learning models. These methods aim to elucidate and disentangle predictions from explanations, requiring only the input and output of the ML model, regardless of its internal structures, as long as the features and model outputs align with the chosen explanation approach [12, 23, 27, 39, 52]. It is worth highlighting that the primary advantage of model-agnostic methods lies in their adaptability to various models. However, despite their convenience, model-agnostic interpretability techniques lack access to specific model parameters, such as weights or structural details [37, 52, 53]. Consequently, they often rely on surrogate models or other approximations that can potentially compromise the accuracy of the explanations they provide [7]. Essentially, model-agnostic explanation methods are post hoc models, as they are applied after the training of the machine learning model [7, 8, 12, 37, 39].
- *Model-Specific methods*: unlike model-agnostic methods, which provide the flexibility to apply across various model architectures, model-specific methods are tailored to explain particular model architecture. As a result, they are limited to a specific class of models, restricting their application solely to those particular model architectures [27, 37, 52, 54]. However, model-specific techniques operate by directly engaging with the model to be interpreted. They delve into the particular characteristics or architecture of the machine learning model, thereby offering comprehensive interpretability that may surpass what is achievable with model-agnostic methods. Therefore, model-specific approaches can yield more precise and detailed explanations [12, 37, 52]. In summary, while possessing more limited applicability, model-specific methods compensate for this constraint by providing deeper insights into the inner workings of the model [52, 55].

3. ***Explainability scope:*** this classification methodology endeavors to categorize XAI methods into global, local, and semi-local explanations according to the explainability scope they offer. These categories seek to understand the entirety of model behavior, the justification behind individual predictions, or a fusion of both aspects.

    - *Global*: global explanation aims to provide a comprehensive understanding of the machine learning model based on a holistic view of its features in a whole data dataset. It focuses on the fundamental aspects of systems and models, encompassing the model's inherent characteristics, behavior, and performance, while also elucidating the comprehensive logic and reasoning leading to diverse potential outcomes [7, 9, 12, 37, 56]. Global interpretation aids decision-makers in attaining a macro-level comprehension of the machine

learning model, including identification of the most influential input features [12].
- *Local*: local explanation in XAI provides clear insights into why a model makes specific decisions or predictions for individual data inputs. This approach focuses on the unique features and characteristics influencing each particular instance rather than the model's behavior as a whole [9, 12, 37, 56, 57]. Such targeted explanations are essential for understanding and evaluating the model's reliability and trustworthiness in real-world applications [57]. This detailed, instance-specific interpretability enhances the transparency and accountability of AI systems.
- *Semi-local*: semi-local explanation in XAI represents a middle ground between local and global explanations. It offers insights into the decision-making process of a model not just for individual instances (as in local explanations) but also for groups of data or on a more global scale. This approach enables a nuanced understanding that captures the reasons behind particular predictions and broader trends or patterns applicable to a dataset subset or category. Such explanations are precious for interpreting complex models where understanding individual and collective behaviors is crucial.

4. ***XAI mechanism inside:*** this classification methodology categorizes XAI methods based on the mechanism inside of XAI techniques. Generally, it can be split into the backpropagation-based and perturbation-based methods.

   - *Backpropagation-based method*: it primarily computes the input characteristics' score based on the model's forward and backward passes. These methods, which include propagation mechanisms and gradient-based approaches, assess how to input characteristics influence the model's predictions. However, their results are not always directly tied to changes in the output, offering insights into the model's internal decision-making processes [34].
   - *Perturbation-based methods*: it involves evaluating the importance of input features by altering, masking, or removing them and observing the impact on the model's output. This is done by comparing the model's predictions on the modified input with those on the original. While these methods offer a direct way to understand the marginal effect of specific features, they can be time-consuming, especially as the number of features increases [34, 58].

5. ***AI execution task:*** this classification methodology categorizes XAI techniques based on the AI model's specific tasks, such as clustering, regression, and classification purposes. Specifically, clustering methods focus on grouping similar data points; regression methods are used to predict continuous outcomes and classification methods are designed to categorize data into predefined classes.
6. ***Data input nature:*** regarding the data input types, XAI approaches can be tailored according to the nature of the data being processed, such as graphs, text, vision, and tabular data. Each data type presents unique challenges and characteristics that necessitate distinct explainability approaches. For instance, vision

data might require techniques to interpret and explain image recognition processes. In contrast, tabular data, mainly involving sequential or time-dependent aspects, would benefit from methods capable of unraveling patterns using Long Short-Term Memory (LSTM) networks or similar approaches. This categorization helps apply the most effective XAI methods and advance the understanding of AI models in diverse application domains.

Exploring distinctions and interrelationships in XAI categorization perspectives In the realm of XAI, comprehending the distinctions and interrelationships among diverse explanation methods is crucial, as it can alleviate potential confusion. The classification of XAI methods is not absolute and varies extensively based on method components, resulting in distinct categories, some of which may overlap [37]. Figure 6 visually depicts these relationships, enhancing clarity and facilitating a deeper comprehension of how these methodologies intersect and diverge. These relationships can be described as follows:

- *AI model dependency and AI Stage:* model-agnostic explanations are inherently post-hoc, meaning they are applied after model development and are not reliant on the model's internal structure [56]. Conversely, ad-hoc methods are model-specific, integrated into the model's architecture during development to align with its unique characteristics [7, 56]. For example, linear regression and decision trees are considered ad-hoc due to their built-in explainability. At the same time, approaches like Grad-CAM, CAM, and DeepLIFT are categorized as post-hoc methods for interpreting models after the training phase.

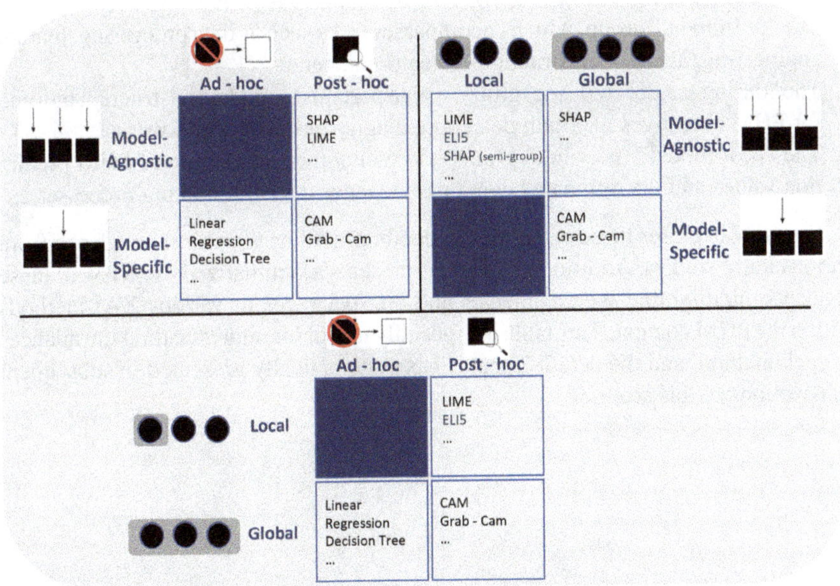

Fig. 6 Relation and difference between groups in XAI

- *Explainability scope and AI Stage:* model-specific methods generally offer global explainability, as they aim to provide an overarching understanding of the model's structure [27]. Conversely, model-agnostic methods can cater to either local or global explanations. For instance, LIME and ELI5 are model-agnostic tools that typically focus on local explanations. However, some model-agnostic methods like SHAP can provide global explanations. It's important to note that SHAP falls into the semi-local category, meaning it can explain local and global aspects of a model's decisions.
- *Explainability scope and AI model dependency:* an ad-hoc method is a model-specific approach, thereby aligning with the characteristics of a taxonomy type, often resulting in the provision of a global explanation. On the other hand, post-hoc methods can cater to both local and global explanations. For instance, tools like LIME and ELI5 are geared towards local explanations, offering insights into individual decisions made by the model. Conversely, methods such as CAM and Grad-CAM are oriented toward providing a global explanation, shedding light on the overall decision-making process of the model.

4 Taxonomy of XAI in PHM

In light of the diverse XAI categorization perspectives discussed earlier, this section focuses on the field of PHM. We propose a taxonomy aligned with PHM stages, classifying XAI methods into three primary groups:

1. The "pre-model" group, which encompasses tasks such as data processing, feature engineering, and the construction of health indicators.
2. The "during-model" (ad-hoc) group, which pertains to model construction tailored for PHM objectives like fault detection, diagnostics, and prognostics.
3. The "post-model" (post-hoc) group, addressing the explanation of PHM prediction values and the rationale behind maintenance decision-making processes.

Figures 7 explores these categories in depth, providing supporting evidence from various PHM studies. Additionally, Table 1 presents a summarized overview of these concepts, highlighting the advantages and disadvantages of various XAI methods within the PHM context. This table is especially useful for understanding the nuances of each method, and the detailed analysis is systematically presented in subsequent sub-sections of this section.

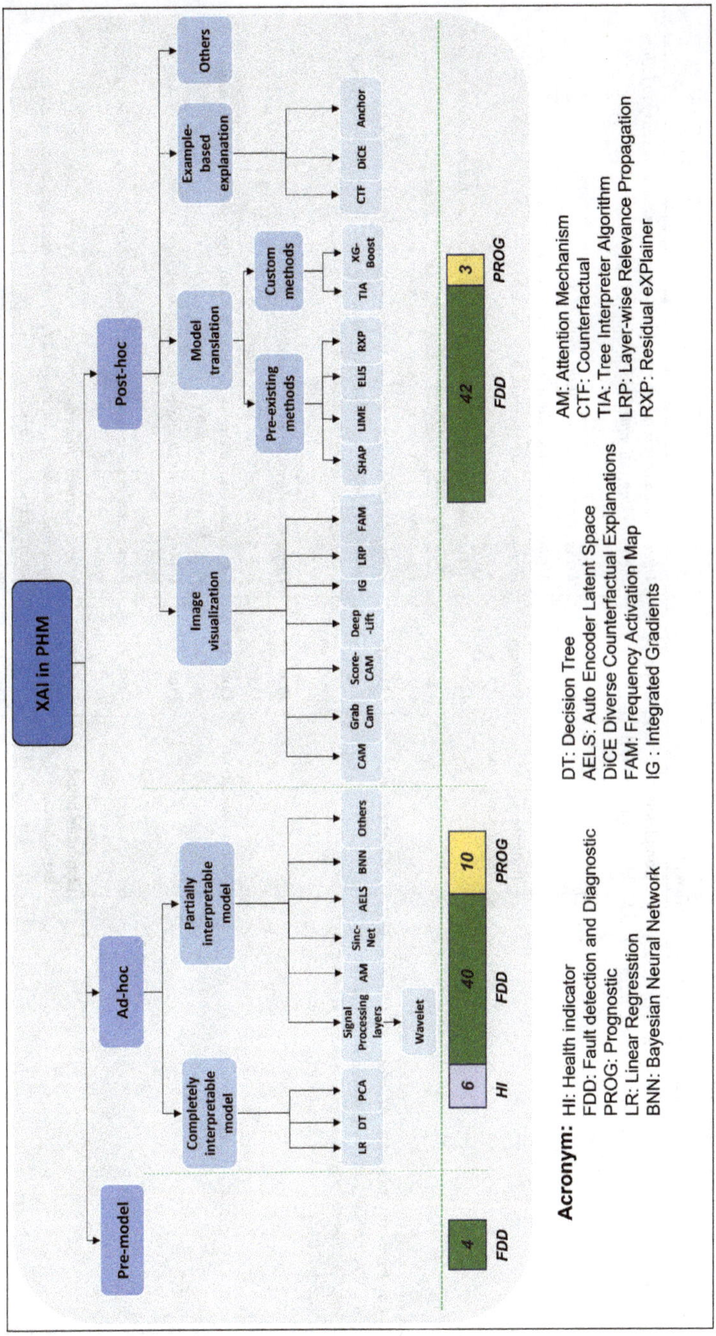

Fig. 7 Categories of XAI methods in PHM

Table 1 Synthesizing XAI methods in PHM

Pre-model	Concept	*Involves exploring, understanding, and describing the data before model construction*		
	Pros	*Understands the data, help easier give the strategies to build a model*		
	Cons	*Focuses on the data only, the model is still a black box*		
Ad-hoc	*Denotes models that have transparent architecture. The interpretability is constructed during the stage of building or training the model*	Completely interpretable model	Concept	*Offers complete transparency by simple structures, and are easy to follow their working rationale behind*
			Pros	*Brings transparency to all the steps of the model*
			Cons	*Methods are simple which leads to low accuracy*
		Partially interpretable model	Concept	*Brings partial interpretability for the model*
			Pros	*Brings higher accuracy through it's complexity, can balance the accuracy and complexity of AI model*
			Cons	*The model is only interpreted in some parts, not entire*
Post-hoc	*Can be applied to explain AI models that have intricate architecture and lack interpretability. This method can be applied after the stage the model is built, often employing external tools or techniques for explanation*	Visualization	Concept	*Creates explanations from gradient-weighted methods, model propagation techniques aim to visualize their representations to explore and analyze the hidden patterns inside a neural unit*
			Pros	*Helps explainability more visually*
			Cons	*Explainability for the CNN model only*
		Model translation	Concept	*Creates a surrogate model aiming to mimic the behaviors of the original mode for explainability*
			Pros	*Brings flexibility thanks to the ability to explain different types of data and models*
			Cons	*It only mimics the behavior of the black box model, it's not the black box itself*
		Example-based explaination	Concept	*Selects instances from the dataset or create new instances, identify influential instances for the model parameters or output, or create a counterfactual explanation*
			Pros	*Shows the direct action need to change to achieve the expected results*
			Cons	*There are various ways to reach the same outcome*

4.1 Pre-model Group

Pre-model involves exploring, understanding, and describing the data before model construction. In the task of bearing fault diagnosis, Brusa et al. [59] utilizes STFT to extract spectrograms, which are then projected onto the space of eigen-spectrograms derived from Randomized Principal Component Analysis (RPCA) for interpretability analysis. By using Logical Analysis of Data, Ragab et al. [60, 61] built a model to detect interpretable patterns, which are linked to underlying physical phenomena, aiming to diagnose the faults in chemical processes. In another work, Ma et al. [62] introduced a novel sparsity-constrained GAN (SC-GAN) for generating stable raw vibration signals. The model's hidden layers enable the learning of distinct features, capturing crucial vibration frequencies from time-domain signals, enhancing interpretability, and facilitating fault diagnosis through a subsequent DNN model. In addition, some signal processing techniques can be integrated into the model to make it more transparent. For instance, the authors in [63, 64] introduced a Wavelet Denoising Network for effectively handling large noise in signal processing, leveraging spare information between sensors, and producing interpretable feature maps. Liu et al. [65] combined Morlet wavelet filters with traditional CNN layers for fault diagnosis in rolling bearings, demonstrating interpretability through visualization at each layer [65]. Similar strategies were employed by others in their models, as mentioned in Shang et al. and Wang et al. [49, 66].

Despite the advantages of pre-model methods in PHM, such as aiding in data comprehension and providing strategies for AI model development, this category still encounters limitations while it focuses on the data only and the model is still a black box. Authors often forgo this step while achieving high accuracy using AI models, leading XAI research to focus on two remaining methods, ad-hoc and post-hoc.

4.2 Ad-Hoc Group

As previously discussed, the ad-hoc method refers to models with transparent architectures, where interpretability is established during the model's construction or training phase. This method inherently possesses interpretability without relying on external tools for explanation. It encompasses both completely interpretable models and partially interpretable models.

- ***Completely interpretable models***: models in this category, such as linear regression and decision trees, have simple and transparent structures, making them easily understandable but potentially less accurate than complex models [7, 27, 40]. In the context of PHM tasks, Mayilvahanan et al. [67] employed simple algorithms like linear regression, logistic regression, and tree algorithms to predict lion battery degradation, comparing transparent models for this data type. Similarly,

Langone et al. (2022) utilized logistic regression with elastic net regularization to predict failure detection probabilities in high-pressure plunger pumps, employing time series representations for interpretability. Matzka et al. [68] predicted and explained tool wear, heat dissipation failure, and power failure in the AI4I dataset using a decision tree algorithm and normalized feature deviation.
In general, completely interpretable models present significant advantages by offering complete transparency through a simple structure, making it easy to understand the mechanisms within the model. However, there is a notable drawback: the simplicity of the structure often leads to lower accuracy. As a result, the application of this method in the field of PHM remains somewhat restricted.

- ***Partially interpretable models***: methods in this category offer partial interpretability by focusing on specific aspects of the model, providing insight into select components without achieving complete model transparency. While they enhance user understanding in targeted areas, these models are task-specific and may lack adaptability for different purposes.
One of the most widely used partially interpretable models in the PHM field is attention-based NN networks. While some literature categorizes them as post-hoc methods, we classify them as "ad-hoc" due to their inherent transparency. Indeed, they naturally reveal information flow maps by assigning weights to inputs or internal features, and visualization tools are solely used to make this transparency visible [10, 57]. For instance, Balakir et al. [69] employ an attention-based recurrent neural network to predict hard disk drive faults and use heatmap visualization for explainability. Su et al. [43] utilize an attention mechanism with GRU and fully connected layers for wind turbine gearbox fault prediction. Li et al. [70] propose variational attention with transformers for interpreting vibration signals from rolling bearings and bevel gears. Xie et al. [71] combine mechanism attention with a DNN model to predict the remaining useful life of fuel cells and determine feature importance. Similarly, the works of Zhang et al. [44, 72, 73] also incorporate interpretable attention mechanisms.
Additionally, partially interpretable models often employ autoencoder latent spaces to explain RUL prediction models. For instance, Costa et al. used latent space representation to construct graphical maps for deterioration patterns [46]. Remadna et al. [47] reduced data dimensionality with an autoencoder, creating 2D visualizations while Jing et al. [45] employed a Transformer Encoder-based model to extract local features [45]. Similarly, Zhang et al. [74] used an LSTM Autoencoder Decoder model for RUL prediction and leveraged the latent space post-autoencoder for interpretation [74].
In addition to prognostic purposes, numerous studies propose partially interpretable models for fault detection and diagnostic tasks. For instance, Abid et al. [48] introduced Deep SincNet for fault diagnosis in induction motors, while Chen et al. [75] incorporated Cosine and Sine blocks for Short-term Fourier transform and Multi-scale Split Dual Calibration Network in rotating machinery fault diagnosis [48, 75].

Furthermore, various other approaches enhance partial interpretability, including Bayesian inference in Bayesian Neural Networks (BNN), path graph construction, integration of physical factors, and knowledge-based models, as demonstrated in studies by Mao et al., Yang et al., Kraus et al., Wang et al., Yong et al., Zhao et al., Xu et al., Yang et al., and Yu et al. [24, 51, 66, 76–83]. These diverse techniques enrich the interpretability within their respective models, highlighting the variety within this taxonomy.

4.3 Post-Hoc

Post-hoc methods are used to explain the behavior of black-box models and are applied after the model has been trained. Based on the analysis from the PHM perspective, methods can be categorized into four main groups: visualization, model translation, example-based explanation, and other approaches.

- ***Visualization:*** within this group, the methods are known by various names such as visual explanations, saliency maps, class activation maps, and masking visualizations [5, 7, 10, 37]. These methods aim to visualize neural unit representations to uncover hidden patterns within models. They rely on internal mechanisms (e.g., gradients, activations, weights) for visualization, using techniques like Gradient-weighted methods, Backpropagation, and Forward propagation. Notably, most of these are model-specific methods, including CAM, Grad-CAM, Score-CAM, DeepLIFT, Integrated gradients (IG), Layer-wise Relevance Propagation (LRP), Frequency activation map (FAM), and others.
  In the context of PHM tasks, several methods are employed for explainability. Grezmak et al. [84] utilized continuous wavelet transform and LRP for Gearbox Fault Diagnosis. Agarwal et al. [85] applied LRP to enhance accuracy and analyze chemical processes. LRP was also employed in [86–88]. IG was utilized by Hajgato et al. [21] for identifying signal channels affecting machinery status changes, while Peng et al. [89] used IG for fault detection. CAM and its variants, such as Grab-CAM, Scored CAM, and FAM, have been frequently applied in various PHM studies, offering visual explainability, but primarily for CNN models, imposing certain limitations on their versatility [12, 90–97].
- ***Model translation:*** also known as proxy models, surrogate approaches, or model-based methods, they involve creating a simplified surrogate model, such as linear or decision tree models, to mimic the behavior of the original model for the purpose of explanation [10, 27, 37, 98]. These surrogate models are used to reflect feature importance by perturbing input or internal components and observing their impact on model performance, providing insights into feature relevance [7, 9]. Methods in this model translation category can be grouped into custom methods and pre-existing methods.

  - *Custom methods* include methods where the author builds their own surrogate model to explain their purpose. In the work of Schmetz et al. [99], authors built a tree interpreter to explain for RF algorithm in the task of cutting tool wear prediction. In another study on the FDD task, Chatterjee et al. [100] employed a less complex XGBoost model, which was designed to mimic the prediction model LSTM, in order to extract feature importance from the input data of the wind turbine.
  - *Pre-existing methods* denote methods in which authors leverage available tools for explainability, including common tools belonging to the model-agnostic such as SHAP, LIME, ELI5, RXP, etc. In the realm of PHM, these XAI methods are employed commonly for explainability purposes. Choi et al. [104] used SHAP in the FDD task to identify which factors damage the systems. In another work, Huang et al. [101] utilized DeepSHAP aiming to eliminate specific residual features extracted from their AI model. Additionally, the widespread popularity of the SHAP method is evidenced by its application in various studies, as demonstrated in the works of [102–106], and [107]. Apart from SHAP, some other methods are also found. It is the case of LIME which is used to detect the contribution of features from sensors in the study of Madhikermi et al. [108]. With the same task of fault detection, Bukhsh et al. [109] leveraged LIME for their research as well. Additionly, methods such as ELI5 and RXP are found in the work of [110, 111] for RUL and fault detection tasks, respectively. Notably, some methods in this taxonomy created the model-agnostic visualization through the surrogate model from the perturbation mechanism, which differs from the model-specific visualization mentioned previously.

  While the methods in this group are widely used and offer flexibility for explaining various data types and models, they involve approximating a proxy model. Hence, it is important to note that there are generally no theoretical guarantees that a simple surrogate model faithfully represents a more complex model, and users should be aware of this limitation [7, 27, 32, 98].
- ***Example-based explanation:*** methods in this category elucidate the model's task by either selecting instances from the dataset, generating new instances, pinpointing influential instances for model parameters or output, or crafting counterfactual explanations [98]. Similar to human reasoning, these explanations primarily revolve around identifying representative examples that capture the underlying relationships and correlations within the model under scrutiny [23]. In the domain of PHM, Jakubowski et al. [112] estimated roll wear using equations and incorporated these data into an autoencoder (AE) model to predict rolling wear levels. They employed counterfactual explanations to elucidate the model's decision-making process without divulging the complete algorithm logic. In another study by Harinarayan et al. [26], a framework with two processes was proposed, and explainability was integrated into the second process using XGBoost in combination with counterfactual explanations, specifically Diverse Counterfactual Explanations (DiCE), to enhance model accuracy. Generally, within the PHM domain,

the example-based explanation methods primarily rely on counterfactual techniques for explanation, although the number of studies in this category for PHM remains limited.

5 Quantitative Analysis of XAI Trend in PHM

The integration of XAI into various PHM tasks plays a pivotal role in establishing trust among developers and stakeholders. This trust is essential for the practical implementation of PHM solutions. In this light, a quantitative analysis of XAI's impact on PHM is discussed in this section. Leveraging statistical data derived from our methodology, as discussed in Sect. 2.1, we comprehensively analyze the trends and contributions of XAI within each distinct PHM subgroup.

Figure 8 presents a pie chart illustrating the percentage breakdown of XAI integration in PHM tasks. It highlights a predominant focus on the fault detection and diagnostic (FDD) task in literature. Notably, within this specific realm of PHM, 82% of papers leverage XAI techniques, underscoring the crucial role of explanatory approaches in illuminating and comprehending FDD models. A smaller gray slice, making up 12% of the chart, represents the prognostic (PROG) task. The smallest segment, in blue, represents the health indicator (HI) construction task with 6% of the total integration. This chart suggests an unbalanced distribution of XAI applications across different PHM tasks. The dominance of FDD in XAI-integrated PHM tasks results from its extensive development over time. This has led to its widespread application and a deeper body of research. Besides, this statistical analysis has revealed a significant research gap in the explanation of HIs, which are essential for monitoring the condition of systems. Despite their importance, HIs have not been studied as extensively in the context of XAI, pointing to a vital area for future research to ensure effective system health management.

Figure 9 delves into the distribution of each XAI group (Pre-model, Ad-hoc, and Post-hoc) within the respective PHM tasks. We found that the application of the pre-model method appears limited across all three PHM tasks. Researchers often

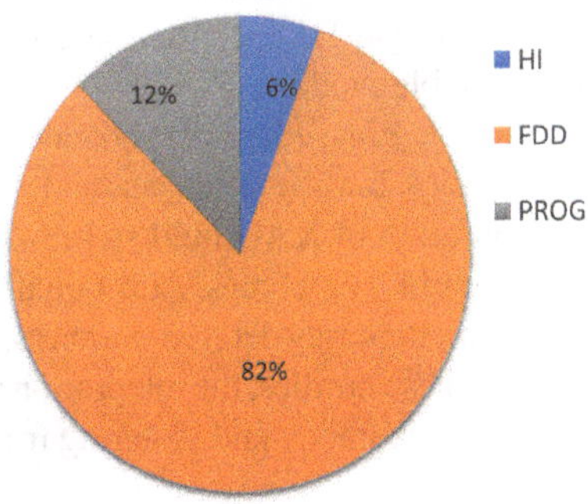

Fig. 8 Percentage breakdown of XAI integration in PHM tasks

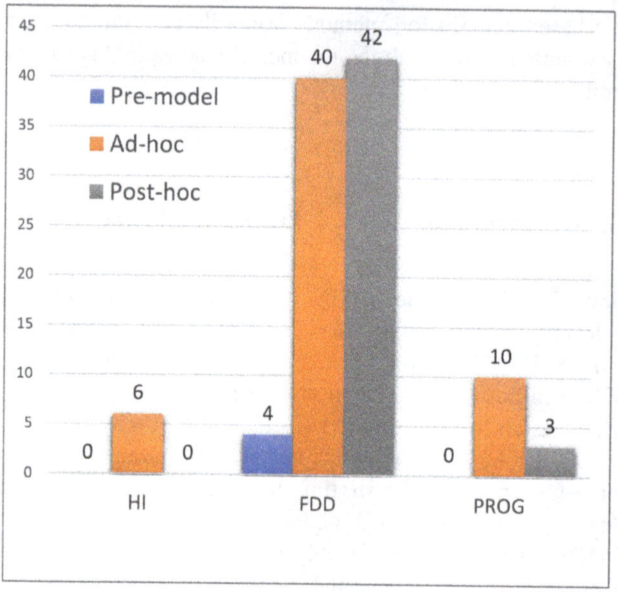

Fig. 9 Distribution of each XAI method group within the respective PHM tasks

bypass this stage in favor of direct accuracy gains through AI models, thus shifting the research focus to ad-hoc and post-hoc methods. These latter methods are predominantly applied within FDD, reflecting a research concentration that underscores their importance in this area. However, the use of post-hoc methods in HI construction and prognostics remains limited, signaling an area ripe for research expansion to leverage their flexibility and enhance these tasks' analytical depth.

Next, we delve into the details of XAI methods's distribution within PHM field, shown in Fig. 10. The horizontal bars indicate the number of papers each XAI method is employed within each PHM task, categorized by method type. Again, this highlights the prevalence and maturity of FDD tasks, followed by the prognostics and HI construction groups. The FDD task utilizes a wide range of XAI techniques, including pre-model, fully interpretable models, partially interpretable models, image visualization, model translation, example-based explanations, and other methods. The prognostic and HI construction tasks display a more concentrated use of partially interpretable models.

Overall, Figs. 9 and 10 indicate a modest volume of literature on XAI within PHM tasks, hinting at its nascent stage within the PHM domain. The infrequent incorporation of XAI could stem from its complexity and the subsequent difficulty in interpreting outcomes. Additionally, concerns over the fidelity and inherent biases of XAI models may deter researchers, given the high stakes of PHM decision-making. Despite these hurdles, the integration of XAI in PHM is promising, offering enhanced model transparency and fostering trust among end-users and stakeholders.

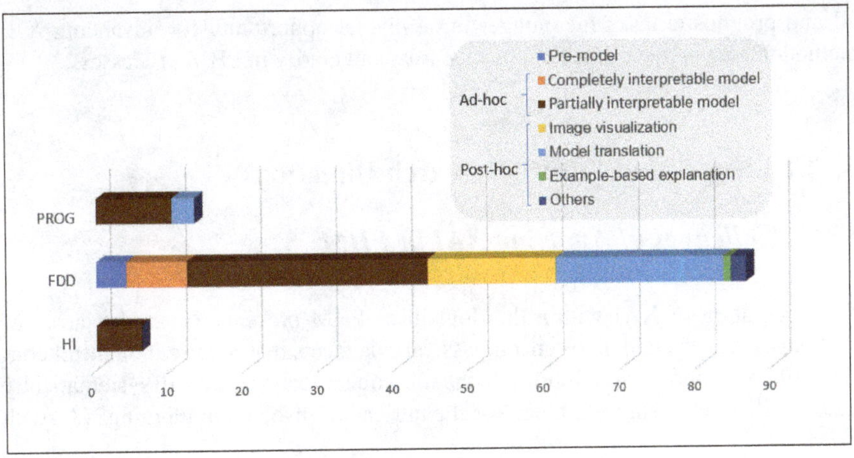

Fig. 10 Distribution of XAI methods used in each PHM task

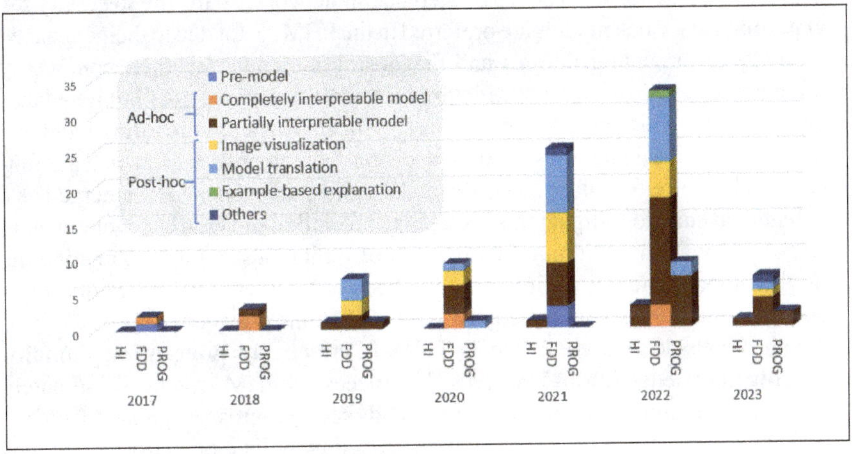

Fig. 11 Development of XAI for PHM each year (until 8/2023)

In literature research, it is encouraging to see that there is a growing interest in applications of XAI in the PHM field, as shown in Fig. 11. This figure demonstrates an exponential increase in research on XAI within the PHM field from 2017 to 2022. It also provides a granular analysis of XAI sub-groups within the primary PHM tasks, showcasing their respective evolution. It finally highlights a surge in adopting partially interpretable models, image visualization techniques, and model translation methods. These methods balance accuracy and explainability with the added benefit of improved reliability due to their self-explanatory characteristics, likely contributing to their rising prominence. Visualization and model translation methods within the post-hoc category are also gaining traction due to their flexibility and intuitive nature. However, while FDD tasks exhibit significant XAI integration,

HI and prognostic tasks lag, suggesting a pivotal opportunity for advancing XAI methodologies in these areas to enhance trust and clarity in PHM processes.

6 Challenges and Future Research Directions

6.1 Challenges of Applying XAI in PHM

The integration of XAI within the domain of PHM presents a complex array of challenges. These challenges encompass various facets that require thoughtful consideration and innovative solutions. In the subsequent sections, we will systematically examine these challenges to elucidate the intricacies involved in merging XAI with PHM.

1. A primary concern is **the difficulty of balancing the intricate nature of advanced AI models**, particularly deep neural networks, **with the necessity for explainability**. Such models are preferred in the PHM sector due to their enhanced accuracy in predicting failures and diagnosing systems. Yet, their complexity often obscures the transparency required for explanation and trust in high-stakes environments. This obscurity presents a significant barrier to providing lucid and understandable explanations for their decision-making processes, which are fundamental to the ethos of XAI. While model-agnostic methods have emerged as a potential avenue to mitigate this issue, they are still in their early developmental stages. They are not without their own set of limitations. By approximating the original model's functionality, these methods can sometimes lead to questions about the fidelity and authenticity of the explanations provided.
2. Another considerable challenge in PHM is **effectively integrating domain-specific knowledge into AI systems.** The success of PHM systems is intimately linked to their application in specialized fields such as aerospace, manufacturing, and the automobile industry, each characterized by unique operational demands and a rich reservoir of expert knowledge. Integrating this technical knowledge into AI models is not only crucial for enhancing their predictive accuracy and relevance but also for ensuring that the explanations generated by these models are meaningful and intelligible to domain experts. This integration demands not just technical skills but also a profound understanding of the specific domain, making it a complex and multi-layered task.
3. For AI models **to deliver accurate predictions and generate meaningful explanations, they necessitate data that are not only high in quality but also pertinent to the specific context.** This becomes **particularly challenging in areas where data may be scarce, noisy, or imbalanced**. Such data quality and availability limitations can severely impair the performance of AI models and, by extension, significantly impact the reliability and clarity of the explanations they offer. Therefore, addressing these data-related issues is essential for developing

robust AI systems in PHM that can provide both precise predictions and transparent decision-making processes.

4. Furthermore, PHM's stakeholders encompass individuals with a **broad spectrum of technical knowledge**. This diversity **necessitates the development of XAI methods capable of delivering explanations tailored to varying expertise levels**. Such explanations must be adaptable, ranging from detailed, technical elucidations for field experts to simplified, intuitive explanations for non-technical personnel. Crafting XAI methods that can seamlessly bridge this communication gap is crucial, ensuring that AI decisions are transparent and understandable to all stakeholders involved in PHM, regardless of their technical background.
5. Additionally, the **implementation of XAI in PHM often demands significant computational resources.** This is particularly true for large-scale, real-time systems with extensive processing needs. Striking an optimal balance between the computational intensity required for thorough XAI applications and maintaining system efficiency presents a tangible challenge. Efficiently integrating comprehensive explanations without overburdening the system's computational capacity is crucial for deploying XAI in PHM, especially in scenarios demanding real-time analysis and decision-making.
6. Another significant challenge in PHM also lies in **developing robust evaluation metrics for XAI.** Given the diverse nature of explanations generated by XAI methods and the inherently subjective nature of user perception, creating standardized evaluation frameworks is complex. It is imperative to establish universally accepted benchmarks that can effectively assess the effectiveness and utility of XAI approaches across different PHM scenarios. This involves not only measuring the accuracy of the explanations but also gauging their understandability and relevance to various user groups. Developing such comprehensive evaluation standards is crucial for advancing the field of XAI in PHM, ensuring that the explanations meet all stakeholders' varied needs and expectations.
7. Furthermore, **the prevalent focus on explaining component-level degradation in isolation must evolve towards more holistic approaches dedicated to system-level explanation**. Understanding and explaining the interactions and collective behaviors of multiple components within a system is vital, especially in the context of system degradation and failure. Grasping these inter-component dynamics is critical to advancing PHM capabilities, as it allows for a more comprehensive understanding of system health and more accurate failure predictions.
8. Lastly, a **notable gap in current research within XAI pertains to the focus on unimodal models, often overlooking the complexities of multimodal learning**. In PHM, multimodal models, which integrate data from diverse sources and modalities, are critical for a comprehensive understanding of complex systems. However, the explainability of these multimodal systems remains under-explored. This lack of focus on multimodal explainability represents a significant shortfall, as it is essential for interpreting and understanding the interplay of various data types and sources. Addressing this gap is crucial for advancing XAI in PHM, ensuring that explanations encompass the full spectrum of data inputs and their interactions within these sophisticated models.

6.2 Future Research Directions in Developing XAI-Assisted PHM

Building upon the challenges elucidated in the previous section, the evolving landscape of XAI in PHM opens up numerous promising avenues for future research endeavors. These avenues are dedicated to surmounting the existing challenges and, in turn, elevating the effectiveness and applicability of the XAI techniques. In this section, we delve into these potential research directions, shedding light on the innovative approaches and methodologies that hold the promise of advancing the integration of XAI within the PHM.

1. A primary focus should be on **developing methods that further balance the complexity of advanced AI models with their explainability**. This entails not only refining existing model-agnostic approaches but also innovating new techniques that provide more transparent and faithful interpretations of complex models like deep neural networks. Additionally, there's a pressing **need for research that integrates domain-specific knowledge more deeply and effectively into AI models**. This requires a concerted effort to understand the nuances of various industries where PHM is applied and to tailor AI systems to meet these unique requirements while ensuring that their explanations remain accessible and relevant to domain experts.
2. Another significant direction for future research is the **enhancement of data quality and accessibility in PHM applications.** Efforts should be directed towards developing methods for effective data curation, noise filtering, and handling imbalanced datasets, thereby ensuring that AI models are trained on data that is not only high-quality but also contextually relevant. This will improve the accuracy of predictive models and the clarity of their explanations.
3. Addressing the diverse needs of stakeholders in PHM is also crucial. Future developments should include **creating adaptive XAI methods capable of delivering customizable explanations**. These should cater to a spectrum of users, from experts requiring detailed technical insights to non-technical personnel needing simplified, intuitive interpretations. Such adaptability in explanations will democratize the understanding of AI decisions across all levels of technical expertise.
4. In terms of computational efficiency, **innovative strategies are needed to optimize the resource demands of XAI in PHM, especially for large-scale and real-time systems**. Research should focus on developing lightweight yet effective XAI methodologies that do not compromise the system's performance while providing comprehensive explanations.
5. The **establishment of robust evaluation metrics for XAI in PHM** represents another critical research direction. This involves formulating standardized, universally accepted benchmarks that can objectively assess the utility and effectiveness of XAI methods, taking into account the accuracy, understandability, and relevance of the explanations to diverse user groups.
6. Moreover, **a shift towards more systemic approaches in understanding and explaining the health and degradation of complex systems** is necessary. Rather

than focusing solely on component-level analysis, research should aim to elucidate the interactions and collective behaviors of multiple components within a system. This holistic view is vital for a more comprehensive understanding of system health and making more accurate predictions of failures.

7. Finally, **addressing the current gap in research on multimodal models in XAI** is essential. Future studies should explore the explainability of these models, which integrate data from various sources and modalities. This will aid in understanding the complex interplays within these systems, ensuring that explanations are holistic and encompass the full spectrum of data inputs and their interactions.

7 Conclusion

Throughout this chapter, we have highlighted the indispensable role of XAI in enhancing PHM systems. Our exploration reveals that the integration of XAI is not merely a technological advancement but a fundamental step towards establishing trust and enabling effective decision-making in critical applications. This paradigm shift is crucial in an era where reliance on AI rapidly grows in complex and high-stakes environments.

In addressing the challenges of standardizing concepts and terms in XAI for PHM, our work contributes significantly to the field by proposing a clear taxonomy. This effort is vital for advancing coherent and unified practices in XAI application, ensuring its benefits are fully realized in PHM systems.

Moreover, we emphasize the future of AI in PHM as more reliable, transparent, and inherently user-friendly. This focus on user-centric design is pivotal in ensuring that AI technologies are not only efficient but also accessible and intuitive for end-users. By enhancing the user experience, we foresee a notable improvement in system performance and a broader acceptance of AI in various critical applications.

Reflecting on our findings, it's clear that integrating XAI into PHM has profound implications for the future of AI in various industries. This integration promises incremental improvements and a transformative leap in how AI is perceived, utilized, and trusted. It marks a significant step towards an era where AI is not just a tool but a collaborator capable of providing clear, understandable insights in complex decision-making scenarios.

In summary, this chapter lays a solid foundation for future innovations in PHM solutions, where XAI plays a central role. It sets the stage for a future where AI's potential is fully harnessed in PHM, leading to systems that are not just technologically superior but also aligned with the needs and expectations of their users.

Acknowledgements The authors acknowledge the support of the French Agence Nationale de la Recherche (ANR), under grant ANR-22-CE10-0011-01 (project X-IMS).

References

1. Gouriveau R, Medjaher K, Zerhouni N (2016) From prognostics and health systems management to predictive maintenance 1: monitoring and prognostics . Wiley
2. Nguyen KT, Medjaher K, Tran DT (2023) A review of artificial intelligence methods for engineering prognostics and health management with implementation guidelines. Artif Intell Rev 56:3659–3709
3. Atamuradov V, Medjaher K, Dersin P, Lamoureux B, Zerhouni N (2017) Prognostics and health management for maintenance practitioners-review, implementation and tools evaluation. Int J Prognost Health Manag 8:1–31
4. Tsui KL, Chen N, Zhou Q, Hai Y, Wang W et al (2015) Prognostics and health management: a review on data driven approaches. Mathematical Problems in Engineering
5. Ibrahim R, Shafiq MO (2023) Explainable convolutional neural networks: a taxonomy, review, and future directions. ACM Comput Surv 55:1–37
6. Alizadeh F, Esau M, Stevens G, Cassens L (2020) explainable ai: take one step back, move two steps forward
7. Adadi A, Berrada M (2018) Peeking inside the black-box: a survey on explainable artificial intelligence (xai). IEEE Access 6:52138–52160
8. Yang G, Ye Q, Xia J (2022) Unbox the black-box for the medical explainable AI via multi-modal and multi-centre data fusion: a mini-review, two showcases and beyond. Inf Fusion 77:29–52
9. Li X-H, Cao CC, Shi Y, Bai W, Gao H, Qiu L, Wang C, Gao Y, Zhang S, Xue X et al (2020) A survey of data-driven and knowledge-aware explainable AI. IEEE Trans Knowl Data Eng 34:29–49
10. Gilpin LH, Bau D, Yuan BZ, Bajwa A, Specter M, Kagal L (2018) Explaining explanations: An overview of interpretability of machine learning. In: 2018 IEEE 5th international conference on data science and advanced analytics (DSAA), IEEE, pp 80–89
11. Sharma J, Mittal ML, Soni G (2022) Condition-based maintenance using machine learning and role of interpretability: a review. Int J Syst Assur Eng Manag 1–16
12. Chen Z, Xiao F, Guo F, Yan J (2023) Interpretable machine learning for building energy management: a state-of-the-art review. Adv Appl Energy 100123
13. Carter A, Imtiaz S, Naterer G (2023) Review of interpretable machine learning for process industries. Process Saf Environ Prot 170:647–659
14. Yepmo V, Smits G, Pivert O (2022) Anomaly explanation: a review. Data & Knowl Eng 137:101946
15. Cheng X, Chaw JK, Goh KM, Ting TT, Sahrani S, Ahmad MN, Abdul Kadir R, Ang MC (2022) Systematic literature review on visual analytics of predictive maintenance in the manufacturing industry. Sensors 22:6321
16. Tritscher J, Krause A, Hotho A (2023) Feature relevance xai in anomaly detection: reviewing approaches and challenges. Front Artif Intel 6:1099521
17. Sutthithatip S, Perinpanayagam S, Aslam S, Wileman A (2021) Explainable ai in aerospace for enhanced system performance. In: IEEE/AIAA 40th digital avionics systems conference (DASC). IEEE, pp 1–7
18. Antwarg L, Miller RM, Shapira B, Rokach L (2021) Explaining anomalies detected by autoencoders using shapley additive explanations. Expert Syst with Appl 186:115736
19. Carletti M, Terzi M, Susto GA (2023) Interpretable anomaly detection with diffi: depth-based feature importance of isolation forest. Eng Appl Artif Intell 119:105730
20. Amarasinghe K, Kenney K, Manic M (2018) Toward explainable deep neural network based anomaly detection. In: 11th international conference on human system interaction (HSI). IEEE, pp 311–317
21. Hajgató G, Wéber R, Szilágyi B, Tóthpál B, Gyires-Tóth B, Hős C (2022) Predmax: Predictive maintenance with explainable deep convolutional autoencoders. Adv Eng Inf 54:101778
22. Ferraro A, Galli A, Moscato V, Sperlì G (2023) Evaluating explainable artificial intelligence tools for hard disk drive predictive maintenance. Artif Intell Rev 56:7279–7314

23. Arrieta AB, Díaz-Rodríguez N, Del Ser J, Bennetot A, Tabik S, Barbado A, García S, Gil-López S, Molina D, Benjamins R et al (2020) Explainable artificial intelligence (xai): Concepts, taxonomies, opportunities and challenges toward responsible AI. Inf Fusion 58:82–115
24. Kraus M, Feuerriegel S (2019) Forecasting remaining useful life: Interpretable deep learning approach via variational bayesian inferences. Decis Support Syst 125:113100
25. Barbado A, Corcho Ó (2022) Interpretable machine learning models for predicting and explaining vehicle fuel consumption anomalies. Eng Appl Artif Intell 115:105222
26. Harinarayan RRA, Shalinie SM (2022) Xfddc: explainable fault detection diagnosis and correction framework for chemical process systems. Process Saf Environ Prot 165:463–474
27. Angelov PP, Soares EA, Jiang R, Arnold NI, Atkinson PM (2021) Explainable artificial intelligence: an analytical review, Wiley Interdisciplinary Reviews. Data Min Knowl Dis 11:e1424
28. Lipton ZC (2018) The mythos of model interpretability: In machine learning, the concept of interpretability is both important and slippery. Queue 16(3):31–57
29. Fernandez A, Herrera F, Cordon O, del Jesus MJ, Marcelloni F (2019) Evolutionary fuzzy systems for explainable artificial intelligence: why, when, what for, and where to? IEEE Comput Intell Mag 14:69–81
30. Gleicher M (2016) A framework for considering comprehensibility in modeling. Big data 4:75–88
31. Craven MW (1996) Extracting comprehensible models from trained neural networks, The University of Wisconsin-Madison
32. Barbado A, Corcho Ó, Benjamins R (2022) Rule extraction in unsupervised anomaly detection for model explainability: application to oneclass svm. Expert Syst Appl 189:116100
33. Montavon G, Samek W, Müller K-R (2018) Methods for interpreting and understanding deep neural networks. Digital Signal Process 73:1–15
34. Ding W, Abdel-Basset M, Hawash H, Ali AM (2022) Explainability of artificial intelligence methods, applications and challenges: a comprehensive survey. Inf Sci
35. Thampi A (2022) Interpretable AI: Building explainable machine learning systems, Simon and Schuster
36. Joshi G, Walambe R, Kotecha K (2021) A review on explainability in multimodal deep neural nets. IEEE Access 9:59800–59821
37. Ahmed I, Jeon G, Piccialli F (2022) From artificial intelligence to explainable artificial intelligence in industry 4.0: a survey on what, how, and where. IEEE Trans Ind Inf 18:5031–5042
38. Kharal A (2020) Explainable artificial intelligence based fault diagnosis and insight harvesting for steel plates manufacturing. arXiv:2008.04448
39. Huang Z, Wu Y (2022) A survey on explainable anomaly detection for industrial internet of things. In: 2022 IEEE conference on dependable and secure computing (DSC). IEEE, pp 1–9
40. Speith T (2022) A review of taxonomies of explainable artificial intelligence (XAI) methods. In: Proceedings of the 2022 ACM conference on fairness, accountability, and transparency, pp 2239–2250
41. An B, Wang S, Qin F, Zhao Z, Yan R, Chen X (2023) Adversarial algorithm unrolling network for interpretable mechanical anomaly detection. IEEE Trans Neural Netw Learn Syst
42. Kamath U, Liu J (2021) Explainable artificial intelligence: an introduction to interpretable machine learning. Springer
43. Su X, Shan Y, Li C, Mi Y, Fu Y, Dong Z (2022) Spatial-temporal attention and GRU based interpretable condition monitoring of offshore wind turbine gearboxes. IET Renew Power Gener 16:402–415
44. Tang J, Zheng G, Wei C, Huang W, Ding X (2022) Signal-transformer: a robust and interpretable method for rotating machinery intelligent fault diagnosis under variable operating conditions. IEEE Trans Instrum Meas 71:1–11
45. Jing T, Zheng P, Xia L, Liu T (2022) Transformer-based hierarchical latent space VAE for interpretable remaining useful life prediction. Adv Eng Inf 54:101781
46. Costa N, Sánchez L (2022) Variational encoding approach for interpretable assessment of remaining useful life estimation. Reliab Eng Syst Saf 222:108353

47. Remadna I, Terrissa LS, Al Masry Z, Zerhouni N (2022) Rul prediction using a fusion of attention-based convolutional variational autoencoder and ensemble learning classifier. IEEE Trans Reliab 72:06–124
48. Abid FB, Sallem M, Braham A (2019) Robust interpretable deep learning for intelligent fault diagnosis of induction motors. IEEE Trans Instrum Meas 69:3506–3515
49. Shang Z, Zhao Z, Yan R (2023) Denoising fault-aware wavelet network: A signal processing informed neural network for fault diagnosis. Chin J Mech Eng 36:9
50. An B, Wang S, Zhao Z, Qin F, Yan R, Chen X (2022) Interpretable neural network via algorithm unrolling for mechanical fault diagnosis. IEEE Trans Instrum Meas 71:1–11
51. Yong BX, Brintrup A (2022) Coalitional bayesian autoencoders: Towards explainable unsupervised deep learning with applications to condition monitoring under covariate shift. Appl Soft Comput 123:108912
52. Vollert S, Atzmueller M, Theissler A (2021) Interpretable machine learning: a brief survey from the predictive maintenance perspective. In: 26th IEEE international conference on emerging technologies and factory automation (ETFA). IEEE, pp 01–08
53. Došilović FK, Brčić M, Hlupić N (2018) Explainable artificial intelligence: a survey. In: 41st International convention on information and communication technology, electronics and microelectronics (MIPRO). IEEE, pp 0210–0215
54. Bach S, Binder A, Montavon G, Klauschen F, Müller K-R, Samek W (2015) On pixel-wise explanations for non-linear classifier decisions by layer-wise relevance propagation. PloS one 10:e0130140
55. Du M, Liu N, Hu X (2019) Techniques for interpretable machine learning. Commun ACM 63:68–77
56. Van der Velden BH, Kuijf HJ, Gilhuijs KG, Viergever MA (2022) Explainable artificial intelligence (XAI) in deep learning-based medical image analysis. Med Image Anal 79:102470
57. Theissler A, Spinnato F, Schlegel U, Guidotti R (2022) Explainable AI for time series classification: a review, taxonomy and research directions. IEEE Access
58. Bygrave LA (2020) Article 22 automated individual decision-making, including profiling. In: The EU general data protection regulation (GDPR), Oxford University Press
59. Brusa E, Delprete C, Di Maggio LG (2023) Eigen-spectrograms: an interpretable feature space for bearing fault diagnosis based on artificial intelligence and image processing. Mech Adv Mater Struct 30:4639–4651
60. Ragab A, El-koujok M, Amazouz M, Yacout S (2017) Fault detection and diagnosis in the tennessee eastman process using interpretable knowledge discovery. In: annual reliability and maintainability symposium (RAMS). IEEE, pp 1–7
61. Ragab A, El-Koujok M, Poulin B, Amazouz M, Yacout S (2018) Fault diagnosis in industrial chemical processes using interpretable patterns based on logical analysis of data. Expert Syst Appl 95:368–383
62. Ma L, Ding Y, Wang Z, Wang C, Ma J, Lu C (2021) An interpretable data augmentation scheme for machine fault diagnosis based on a sparsity-constrained generative adversarial network. Expert Syst Appl 182:115234
63. Li T, Zhao Z, Sun C, Cheng L, Chen X, Yan R, Gao RX (2021) Waveletkernelnet: an interpretable deep neural network for industrial intelligent diagnosis. IEEE Trans Syst Man Cybernet: Syst 52:2302–2312
64. T. Li, C. Sun, S. Li, Z. Wang, X. Chen, R. Yan, Explainable graph wavelet denoising network for intelligent fault diagnosis, IEEE Transactions on Neural Networks and Learning Systems (2022)
65. Liu C, Ma X, Han T, Shi X, Qin C, Hu S (2022) Ntscatnet: An interpretable convolutional neural network for domain generalization diagnosis across different transmission paths. Measurement 204:112041
66. Wang D, Chen Y, Shen C, Zhong J, Peng Z, Li C (2022) Fully interpretable neural network for locating resonance frequency bands for machine condition monitoring. Mechanical Systems and Signal Processing 168:108673

67. Mayilvahanan KS, Takeuchi KJ, Takeuchi ES, Marschilok AC, West AC (2022) Supervised learning of synthetic big data for li-ion battery degradation diagnosis. Batteries & Supercaps 5:e202100166
68. Matzka S, Explainable artificial intelligence for predictive maintenance applications, in, (2020) third international conference on artificial intelligence for industries (ai4i). IEEE 2020:69–74
69. Balakir A, Yang A, Rosenbaum E, An interpretable predictive model for early detection of hardware failure, in, (2020) IEEE International Reliability Physics Symposium (IRPS). IEEE 2020:1–5
70. Y. Li, Z. Zhou, C. Sun, X. Chen, R. Yan, Variational attention-based interpretable transformer network for rotary machine fault diagnosis, IEEE Transactions on Neural Networks and Learning Systems (2022)
71. Xie R, Li C, Ma R, Xu L, Zhou X (2022) An explainable data-driven framework for fuel cell aging prediction under dynamic condition. IEEE Trans Ind Electron 70:5960–5970
72. Zhang C, Tian X, Zhao Y, Li T, Zhou Y, Zhang X (2022) Causal discovery-based external attention in neural networks for accurate and reliable fault detection and diagnosis of building energy systems. Build Environ 222:109357
73. Liu K, Niri MF, Apachitei G, Lain M, Greenwood D, Marco J (2022) Interpretable machine learning for battery capacities prediction and coating parameters analysis. Control Eng Pract 124:105202
74. Zhang J, Li X, Tian J, Jiang Y, Luo H, Yin S (2023) A variational local weighted deep sub-domain adaptation network for remaining useful life prediction facing cross-domain condition. Reliab Eng Syst Saf 231:108986
75. Chen Y, Zhang D, Ni H, Cheng J, Karimi HR (2023) Multi-scale split dual calibration network with periodic information for interpretable fault diagnosis of rotating machinery. Eng Appl Artif Intell 123:106181
76. Mao W, Liu J, Chen J, Liang X (2022) An interpretable deep transfer learning-based remaining useful life prediction approach for bearings with selective degradation knowledge fusion. IEEE Trans Instrum Meas 71:1–16
77. Yang C, Liu J, Zhou K, Jiang X, Ge M-F, Liu Y (2022) A node-level pathgraph-based bearing remaining useful life prediction method. IEEE Trans Instrum Meas 71:1–10
78. Zhao X, Wang X, Golay MW (2023) Bayesian network-based fault diagnostic system for nuclear power plant assets. Nucl Technol 209:401–418
79. Zhao J, Li Y-G, Sampath S (2023) A hierarchical structure built on physical and data-based information for intelligent aero-engine gas path diagnostics. Appl Energy 332:120520
80. Xu Z, Guo Y, Saleh JH (2022) A physics-informed dynamic deep autoencoder for accurate state-of-health prediction of lithium-ion battery. Neural Comput Appl 34:15997–16017
81. Wang D, Hou B, Yan T, Shen C, Peng Z (2023) New statistical learning perspective for design of a physically interpretable prototypical neural network for machine condition monitoring. Mech Syst Signal Process 188:110041
82. Yang W-T, Reis MS, Borodin V, Juge M, Roussy A (2022) An interpretable unsupervised bayesian network model for fault detection and diagnosis. Control Eng Pract 127:105304
83. Yu J, Liu G (2020) Knowledge extraction and insertion to deep belief network for gearbox fault diagnosis. Knowl-Based Syst 197:105883
84. Grezmak J, Wang P, Sun C, Gao RX (2019) Explainable convolutional neural network for gearbox fault diagnosis. Proc CIRP 80:476–481
85. Agarwal P, Tamer M, Budman H (2021) Explainability: relevance based dynamic deep learning algorithm for fault detection and diagnosis in chemical processes. Comput Chem Eng 154:107467
86. Grezmak J, Zhang J, Wang P, Loparo KA, Gao RX (2019) Interpretable convolutional neural network through layer-wise relevance propagation for machine fault diagnosis. IEEE Sens J 20:3172–3181
87. Wang F, Zhao Z, Zhai Z, Shang Z, Yan R, Chen X (2023) Explainability-driven model improvement for soh estimation of lithium-ion battery. Reliab Eng Syst Saf 232:109046

88. Tod G, Mazaev G, Eryilmaz K, Ompusunggu AP, Hostens E, Van Hoecke S (2019) A convolutional neural network aided physical model improvement for ac solenoid valves diagnosis. In: Prognostics and system health management conference (PHM-Paris). IEEE, pp 223–227
89. Peng P, Zhang Y, Wang H, Zhang H (2022) Towards robust and understandable fault detection and diagnosis using denoising sparse autoencoder and smooth integrated gradients. ISA Trans 125:371–383
90. Liu J, Hou L, Wang X, Zhang R, Sun X, Xu L, Yu Q (2021) Explainable fault diagnosis of gas-liquid separator based on fully convolutional neural network. Comput Chem Eng 155:107535
91. Chen H-Y, Lee C-H (2020) Vibration signals analysis by explainable artificial intelligence (XAI) approach: application on bearing faults diagnosis. IEEE Access 8:134246–134256
92. Lee J, Noh I, Lee J, Lee SW (2021) Development of an explainable fault diagnosis framework based on sensor data imagification: a case study of the robotic spot-welding process. IEEE Trans Ind Inf 18:6895–6904
93. Kim MS, Yun JP, Park P (2021) An explainable neural network for fault diagnosis with a frequency activation map. IEEE Access 9:98962–98972
94. Kim MS, Yun JP, Park P (2022) Deep learning-based explainable fault diagnosis model with an individually grouped 1-d convolution for three-axis vibration signals. IEEE Trans Ind Inf 18:8807–8817
95. Kim MS, Yun JP, Park P (2020) An explainable convolutional neural network for fault diagnosis in linear motion guide. IEEE Trans Ind Inf 17:4036–4045
96. Li G, Yao Q, Fan C, Zhou C, Wu G, Zhou Z, Fang X (2021) An explainable one-dimensional convolutional neural networks based fault diagnosis method for building heating, ventilation and air conditioning systems. Build Environ 203:108057
97. Liu J, Hou L, Zhang R, Sun X, Yu Q, Yang K, Zhang X (2023) Explainable fault diagnosis of oil-gas treatment station based on transfer learning. Energy 262:125258
98. Markus AF, Kors JA, Rijnbeek PR (2021) The role of explainability in creating trustworthy artificial intelligence for health care: a comprehensive survey of the terminology, design choices, and evaluation strategies. J Biomed Inf 113:103655
99. Schmetz A, Vahl C, Zhen Z, Reibert D, Mayer S, Zontar D, Garcke J, Brecher C (2021) Decision support by interpretable machine learning in acoustic emission based cutting tool wear prediction. In: 2021 IEEE international conference on industrial engineering and engineering management (IEEM). IEEE, pp 629–633
100. Chatterjee J, Dethlefs N (2020) Deep learning with knowledge transfer for explainable anomaly prediction in wind turbines. Wind Energy 23:1693–1710
101. Huang Z, Wu Y, Tempini N, Lin H, Yin H (2022) An energy-efficient and trustworthy unsupervised anomaly detection framework (eatu) for iiot. ACM Trans Sensor Netw 18:1–18
102. Park S, Moon J, Hwang E (2020) Explainable anomaly detection for district heating based on shapley additive explanations. In: 2020 international conference on data mining workshops (ICDMW). IEEE, pp 762–765
103. Jakubowski J, Stanisz P, Bobek S, Nalepa J (2021) Explainable anomaly detection for hot-rolling industrial process. In: 2021 IEEE 8th international conference on data science and advanced analytics (DSAA). IEEE, pp 1–10
104. Choi H, Kim D, Kim J, Kim J, Kang P (2022) Explainable anomaly detection framework for predictive maintenance in manufacturing systems. Appl Soft Comput 125:109147
105. Huong TT, Bac TP, Ha KN, Hoang NV, Hoang NX, Hung NT, Tran KP (2022) Federated learning-based explainable anomaly detection for industrial control systems. IEEE Access 10:53854–53872
106. Park JH, Jo HS, Lee SH, Oh SW, Na MG (2022) A reliable intelligent diagnostic assistant for nuclear power plants using explainable artificial intelligence of gru-ae, lightgbm and shap, Nuclear. Eng Technol 54:1271–1287
107. Chowdhury D, Sinha A, Das D (2022) Xai-3dp: diagnosis and understanding faults of 3-d printer with explainable ensemble AI. IEEE Sensors Lett 7:1–4
108. Madhikermi M, Malhi AK, Främling K (2019) Explainable artificial intelligence based heat recycler fault detection in air handling unit. In: Explainable, transparent autonomous agents

and multi-agent systems: first international workshop, EXTRAAMAS 2019, Montreal, QC, Canada, Revised Selected Papers 1. Springer, pp 110–125. Accessed from 13–14 May 2019
109. Bukhsh ZA, Saeed A, Stipanovic I, Doree AG (2019) Predictive maintenance using tree-based classification techniques: a case of railway switches. Transp Res Part C: Emerg Technol 101:35–54
110. Oliveira DF, Vismari LF, Nascimento AM, de Almeida JR, Cugnasca PS, Camargo JB, Almeida L, Gripp R, Neves M (2021) A new interpretable unsupervised anomaly detection method based on residual explanation. IEEE Access 10:1401–1409
111. Serradilla O, Zugasti E, Cernuda C, Aranburu A, de Okariz JR, Zurutuza U (2020) Interpreting remaining useful life estimations combining explainable artificial intelligence and domain knowledge in industrial machinery. In: IEEE international conference on fuzzy systems (FUZZ-IEEE). IEEE, pp 1–8
112. Jakubowski J, Stanisz P, Bobek S, Nalepa GJ (2022) Roll wear prediction in strip cold rolling with physics-informed autoencoder and counterfactual explanations. In: 2022 IEEE 9th international conference on data science and advanced analytics (DSAA). IEEE, pp 1–10

Duc An Nguyen received his Bachelor's degree in Electronics-Telecommunications Engineering from Ho Chi Minh University of Technology in 2019. He continued his academic journey at HCMUT, conducting research and earning a Master's degree in Electronics Engineering in 2021. With a keen interest in practical applications, from 2021 to 2022, he collaborated with an industry company specializing in computer vision technologies. Currently, Duc An is pursuing his Ph.D. at École Nationale d'Ingénieurs de Tarbes, UTTOP, France, focusing on Artificial Intelligence and explainable AI solutions for prognostics and health management of multicomponent systems.

Khanh T. P. Nguyen received the Ph.D. degree in Automatic Control, Production and Robotics from the Ecole Centrale de Nantes, France, in 2012. From 2013 to 2015, she was a Post-Doctoral Fellow with the French Institute of Science and Technology for Transport, Development and Networks (IFSTTAR), Villeneuve d'Ascq, France. From 2016 to 2017, she was an Assistant Professor at the University of Technology of Troyes, France. In September 2017, she joined the École Nationale d'Ingénieurs de Tarbes, UTTOP, France, as an Associate Professor. Her current research interests include prognostics health management, maintenance optimization, artificial intelligence, machine learning, deep learning, and safety assessment of complex systems.

Kamal Medjaher received the Ph.D. degree in control and industrial computing from University of Lille 1, France, in 2005. He was Associate Professor at the National Institute of Mechanics and Microtechnologies, Besançon, France, and member of FEMTO-ST Institute, from 2006 to 2016. He is currently Full Professor at University of Technology Tarbes Occitanie Pyrénées (UTTOP), France. He conducts his research activities within the Production Engineering Laboratory (LGP). His current research interests include prognostics and health management of industrial systems and predictive maintenance. He co-authored more than 50 journal papers, co-supervised 15 Ph.D. students and leaded more than 10 collaborative research projects.

Human Robot Collaboration

Mourad Benoussaad

Abstract The current chapter presents an overview and state of the art in the field of collaborative robotics (commonly called cobotics), where several issues, challenges and contributions in the field are summarized. Therefore, the chapter addresses different aspects of collaborative robotics and its positioning in the context of the whole robotics research. A particular focus is made on physical Human-Robot Interaction (pHRI) where different aspects are addressed according to the level and frequency of physical interaction. It starts from the space sharing and collision avoidance, through the punctual collision, until the permanent physical interaction between humains and robots. Hence, the specific context of physical Human-Robot Collaboration (pHRC), defined as a key task of human-robot object co-manipulation, is deeply addressed. The importance of collaboration quality evaluation is addressed in this chapter as well, and how this field of pHRC can be considered as a part of the human-centered robotics field thanks to its impact on human beings.

1 Introduction

During the last decades, robots were created to replace humans in dangerous or inaccessible areas (nuclear environments, space exploration, disaster and war areas, mini-invasive surgery,...) and also used in several domains to handle repetitive and fastidious tasks. Where robots were usually teleoperated by a distant human (in a master-slave scheme), since a human's decision is needed [1], robots are integrated into the industry for automatizing and improving the production chains thanks to their main qualities such as force capability and high repeatability of accomplishing tasks [2]. This classical approach of industrial robotics is representative of many manipulator robots that perform repetitive tasks for a massive and standard production (e.g., machining, assembly, painting, peak-and-place, ...).

M. Benoussaad (✉)
MAVRICS Research group, Laboratoire Génie de Production (LGP),
Univesity of Technology of Tarbes (UTTOP), 65000 Tarbes, France
e-mail: mourad.benoussaad@uttop.fr

K. P. Tran (ed.), *Artificial Intelligence for Safety and Reliability Engineering*,
Springer Series in Reliability Engineering,
https://doi.org/10.1007/978-3-031-71495-5_7

From the beginning, robotics intends to be more autonomous, with high decision capabilities when interacting in a complex and unexpected environment, and inevitably with humans. However, this purpose has yet to be reached in most robotics applications nowadays, even if it is still targeted, such as for self-driving cars [3].

Alternatively, future robotics should be more interactive and collaborative with humans to accomplish the most complex tasks. Indeed, some tasks require a combination of human capabilities, such as high decision-making and perception, with robot capabilities, such as mechanical stability, precision and time-invariant force. Moreover, human-robot interaction is inevitable for several tasks and contexts, and collaboration between them is needed, which should increase robotics acceptance in different fields. One of the primary current examples where this future robotics needs to be rethought concerns the Industrial New Revolution (Commonly called *Industry 4.0* and *Industry 5.0*).

This future manufacturing intends to be digital, smart, human-centered and sustainable, based on several concepts such as flexible reconfiguration and agility, interconnection and Industrial Internet of Things (IIoT), and communication between entities such as between humans and machines [4, 5]. Even if the framework of future industries is still in a permanent evolution (*Industry 4.0* and *Industry 5.0*), one of its primary purposes is to increase product customization, which opposes massive and standard production done so far in the previous industry. In this case, the principle of a robot working repetitively, by doing the same task in the same way, is no longer appropriate for the concept of customization and the difficulty of regular reprogramming of the robot. Therefore, allowing interaction and making collaboration between humans and robots (commonly called *cobotics*) becomes an exciting approach in this context [6].

On one side, beyond *Industry 4.0*, *Industry 5.0* reinforce aspects of the industry more centered on human [5], which makes the collaboration between human and robot a reality that should take place in the future industry. Conversely, robots' autonomy and decision capabilities may have limits regarding ethics and responsibility aspects, as well as the high level of human expertise usually needed. Therefore, combining human capabilities (intelligence and decision) and robot capabilities (force and precision) may be optimal for achieving several tasks.

However, this human-robot collaboration poses several problems at different levels (technical, scientific, ethical, ...). More specifically, when we target human-robot collaboration, one of the leading concrete questions, is about the role and place of each partner in the collaboration: Should the robot be just a sophisticated tool that responds to human actions without any initiative? Should the robot be a *learner* from humans by integrating artificial intelligence aspects? Should the robot be more of a partner? More proactive by taking the initiative and suggesting actions to reduce the human cognitive load? Answering these essential questions depends strongly on the application and its context. However, it will have consequences on the technical and scientific aspects and issues to be tackled.

In main and challenging cases of future Human-Robot Interaction and Collaboration, robots will become more and more physically closer to active human partners.

According to literature of collaborative robotics, this *physical distance* can be of different levels:

(i) Human and robot may share space by avoiding any physical contact [7, 8],
(ii) human and robot may have a punctual physical contact that will be handled [9],
(iii) a permanent physical contact is required and even targeted [10, 11].

Hence, a generic term of pHRI is commonly used in literature [12], corresponding to *physical Human-Robot Interaction*. However, when this interaction is to accomplish a common task in a collaborative way, such as an object co-manipulation, the term pHRC (physical Human-Robot Collaboration) seems more appropriate. This term is not much used in literature [13, 14] and the reason is probably the fact that the human-robot interaction was firstly linked to works about cobotics, limited to direct physical contact between human and robot, and the collaboration aspect is not enough emphasized, as explicitly by the case of object co-manipulation conjointly between human and robot.

Figure 1 represents and summarizes all these cited aspects, with respect to the large domain of robotics, where the pHRC seems to be a part of pHRI and cobotics in general.

Therefore, the current chapter will introduce aspects of cobotics and cobots (Sect. 2) by introducing the standard/legal aspects. Different levels of physical interaction will be addressed in (Sect. 3), where the specific case of physical collaboration is addressed in (Sect. 4). Addressing physical interaction and collaboration can not be done without considering the human side and perception of the collaboration (Sect. 5), which may favor acceptability aspects of future collaborative robotics. A conclusion of this chapter can be found in (Sect. 6).

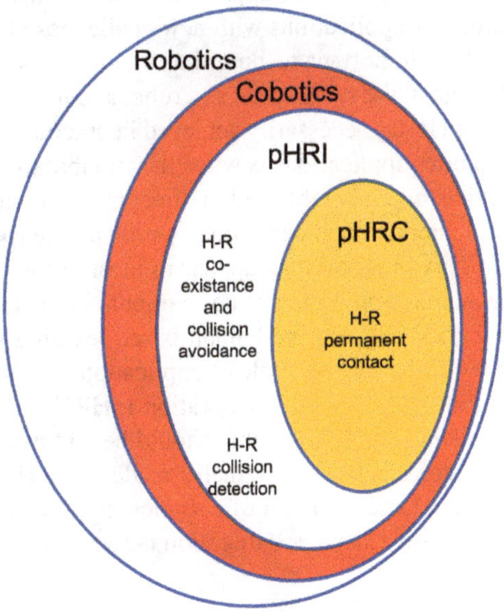

Fig. 1 Positioning of pHRI and pHRC in the large context of robotics and cobotics

2 Robotics, Cobotics and Cobots

Since the beginning of robotics, and particularly in the industrial context, the robot has been introduced to mainly accomplish repetitive tasks where the environment is well structured and somehow controlled. Hence, every component of the environment is known, and even if some dynamical aspect may exist (e.g. pieces to be picked up, localized using a camera on moving traidmail), the action of the robot during execution may be seen as on a static environment. In this case, the robot is controlled based on well-known motion planning, followed by a motion control to track the predefined trajectory in the joint space or the task space [15]. However, in several complex tasks in industry or daily life, the repetitive functioning of the robot with a predefined environment can not be ensured if the environment is not structured enough or presents a lot of uncertainties, and if the robot is in direct physical interaction with humans. Other cases that highlight a limit of classical robotics appears if the complexity of the task requires interaction and collaboration with human, with its intelligence and decision capabilities. In this case, the interaction and collaboration between humans and robots becomes desired.

Therefore, in recent years, new robotic approaches that favor collaboration with human have appeared and is growing. This new scientific and challenging way, which outlines are still not clearly defined regarding classical robotics, is commonly called cobotics (for COllaborative roBOTICS) [16, 17].

As an example, the teleoperation of robots by humans may be seen as a case of cobotics [18]. Moreover, surgery robotics, where the robot is remotely controlled by a human and is in physical interaction with human patients [19], can be seen as cobotics as well. However, these applications have existed since the beginning of robotics, when the concept of cobotics is not introduced yet. Is the cobotics just a definition of some application that existed before? Is it a generalization of some previous applications with new challenges of future robotics?

It is important to note that the cobotics domain commonly involves an active human in the same space as a robot and in direct physical interaction, where collaboration is not necessarily achieved in practice. Therefore, one of the main challenging cobotics applications is when the collaboration is effective, involving a human as an active, unpredictable and in direct physical interaction with the robot, particularly if this interaction is continuous within a time (permanent).

This situation may appear in most of the complex tasks of future robotics, such as human-robot objects co-manipulation [11, 20, 21], assistance and rehabilitation robotics [22], online human-based learning robotics [23] and wearable exoskeletons [24]. In this sample of applications, a continuous physical Human-Robot Interaction (pHRI) and collaboration (pHRC) [12, 25] has been raised with important challenges and issues that should be addressed and remain outstanding. Even if, in literature, pHRI is originally made to reach purposes of collaboration and seems to be a larger domain of cobotics that includes pHRC (see Fig. 1), pHRI does not necessarily mean collaboration between human and robot.

In pHRI context [12], issues related to human safety, comfort and ergonomics were addressed by improving the robot's design in the first place [26]. This improvement consists in making robots lighter, with rounder shapes and their joints more flexible or with a variable flexibility by integrating often a Variable Impedance Actuators (VIA) [27] or Variable Stiffness Actuators (VSA) [27], inspired from co-actuation principle of biology [28, 29]. The purpose of joint flexibility is a way to make the robot more *human-friendly* by making it more and more human-like in terms of mechanical behavior.

Other improvements of collaborative robotics applications require also control strategies through different compliant control techniques [30–32]. For better control through an accurate model, identification protocols and techniques, that consider joint flexibility, online evolution and objects with unknown dynamics and high uncertainties, should be addressed as well [33]. An improvement in robot use and integration has also been introduced, which consists of the integration of all required sensors for more straightforward implementation in an industrial context and a fast robot's programming by demonstration [34].

All these improvements and in-depth redefinition have led to the creation of a new generation of robots, commonly called *cobots* (for COllaborative RoBOTs), which is devoted to working closely with humans for the purpose of the standardization and generalization of collaborative robotics applications. Hence, cobots were equipped with additional sensors and perception's capacities of environment and human and force sensing in contact. In appearance, it seems, therefore, difficult to establish a definition of Cobot and a clear distinction with respect to existing robots used in research, which may consider similar aspects such as environment perception, force control techniques and adaptation capabilities [35, 36].

However, this distinction may be clearly made thanks to *ISO TS 15066* normative standard for Robots and robotic devices—collaborative robots [37, 38]. This standard (ISO TS 15066: 2016) is an extension of generic guidelines of a previous one, ISO 10218: 2011 standard for robots and robotics devices—safety requirements for industrial robots.

The purpose of this ISO TS 15066 standard is to propose guidelines for designers, integrators and users to ensure a collaborative robot system and design collaborative workspaces. This is mainly based on human safety insurance through outcomes of hazard analysis and risk assessment processes and by embedding passive and active safeguards. To design safeguards for collaborative robots, the following *safety requirements* were included in *ISO TS 15066 (Clause 5.5)* [38]:

- Safety-rated monitored stop: Allows automatic or manual-based intervention on robot action.
- Hand guiding: Allows the manipulation of the robot safely by the human and avoids potential hazardous contact.
- Power and force limiting: Establish limiting threshold values for static and quasi-static contact in order to limit the pain sensitivity for parts of the human body.
- Speed and Separation Monitoring (*SSM*): Monitor and maintain a safe separation distance between human and robot [39] and limit robot speed. When the human

share the robot area, the TCP speed is limited to 2400 mm/s if the end-effector exerts maximum pressure with an area of 1 cm^2 of the surface of the hand or finger. This limit is to reduce the impact of possible collision with humans. However, the robot speed may be adaptive according to human detection in predefined safety zones [40, 41]. This requirement is particularly appropriate for collision avoidance techniques [7, 42], but seems incompatible with all pHRI applications, and particularly pHRC where a permanent contact is required.
- Workpiece and end-effector related hazards: Consider risks related to the workpiece, end-effector, and tools attached to the robot manipulator.

Several of these *safety requirements* are not completely independent since they may be combined or one requirement may activate another. However, they allow for the removal of physical barriers between humans and robots in an industrial context. This ISO standard, in reality, refers to collaborative robot systems, commonly called cobots. In conclusion, what is defined as a cobot is a robot that respects the requirements of ISO TS 15066 standards. More details about the ISO TS 15066 standard analysis and its specificity regarding the previous one may be found in [38]. From these safety requirements where forces applied by these robots are limited, an alternative terminology is sometimes used to define cobots as *Force Limited Robots* [43].

It is, however, important to notice that the use of cobots does not necessarily involve a collaborative task and, in several cases, does not involve even pHRI context since the "cobotic" aspect may be used only for easy programming of the robot, where the robot accomplish a predefined and repetitive task in a classical way. Indeed, cobots in today's industry are often used as classical robots without barriers and physical separation, where safety is ensured even if the human does not effectively and continuously share the space.

Therefore, it seems more appropriate, in human-robot collaboration, to talk about collaborative robotics (cobotics) applications than collaborative robots (Cobots). Indeed, Cobotics refers clearly to a collaborative task, while the use of cobots does not mean necessary cobotics application and is often used as a classical robot in today's industry. Moreover, a cobot without appropriate methods and techniques may present limited collaboration perspectives. Hence, integrating a cobot does not ensure collaborative application achievement.

On the other side, a classical robot with additional sensors and appropriate control methods may be used in a cobotic application [44]. Moreover, a robot with a large scale and high payload can be used as a cobot, as is the case of the Aura robot (COMAU, 170 kg of payload) [45]. Hence, a not force-limited robot with appropriate methods can be implemented for a collaborative application, but this may pose some problems in terms of standards and legal aspects. Therefore, in the next of this chapter, when we talk about cobotics, we mean more precisely robotics in a collaborative task and context than a cobot as a turnkey.

Domains of cobotics may be diverse according to the proximity and the level of physical interaction between humans and robots. Hence, as addressed in the next sections, cobotics applications may go from general physical Human-Robot Interaction (pHRI) (Sect. 3), that includes co-existence and space sharing (Sect. 3.1), until a continuous physical Human-Robot collaboration to accomplish more complex tasks (Sect. 4).

3 Physical Human-Robot Interaction (pHRI)

To be more specific, it is important to define the physical Human-Robot Interaction (pHRI) in the general context of Human-Robot Interaction (HRI). Unlike HRI or H-R cooperation [46], where the human and robot do not share the same space even if they interact with each other or collaborate (e.g. teleoperation, split a whole task into subtasks or acting on different times), pHRI implies working in the same shared space, without physical separation, where the physical contact between human and robot is possible and sometimes required. Therefore, a classification of pHRI may be summarized in Fig. 2 where we can see that, from the top to the bottom, increasing the level (frequency) of physical interaction. Hence, it starts from collision avoidance, going through punctual physical contact, until continuous physical contact, which a list of possible situations is listed at the bottom. This classification is largely inspired by previous works in pHRI, as detailed in [13], where the authors described a framework considering collision avoidance, detection, reaction and possible collaboration.

In Fig. 2, different situations of pHRI with continuous (permanent) contact may be classified according to the increasing level of human's leadership (bottom part), where we can thus identify (from the left to the right):

1. *Robot manipulating human*. This concerns, for example, applications of robotic-based physical rehabilitation, where a robot is used to make a repetitive motion for human limbs that is completely passive [47, 48]. This case was addressed in one of our previous works that focused on the robust control of a robot in interaction with a passive human body, represented by a biomechanical model [49]. This

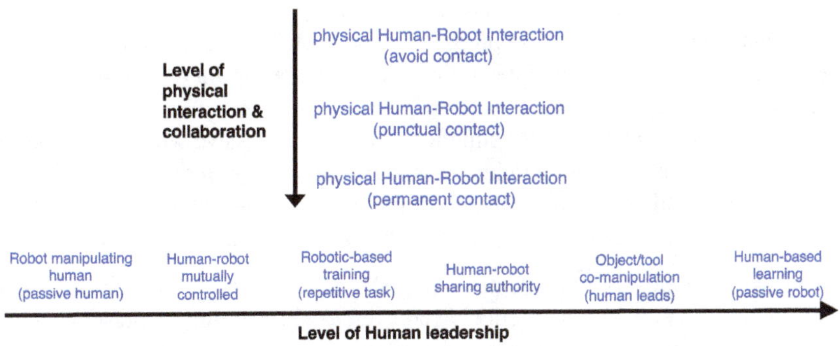

Fig. 2 Levels of pHRI and human's leadership

situation may concern other cases where the human is physically passive, such as an active exoskeleton [50, 51], or surgical robots manipulating organs [52, 53].

2. *Human-robot mutually controlled.* In this type of application, the human and robot are both controlled [54, 55], where the human body can be controlled through Functional Electrical Stimulation (FES) [56, 57]. This concerns particularly the commonly called hybrid exoskeleton [54]. In this case, even if the control of the human body may pose a certain number of issues, related to its high complexity, variability, and uncertainties [56], the problem of coordination between human and robot are extremely reduced since the system may be controlled as a whole.
3. *Robotic-based training.* This case concerns robotic solutions that may be used for human physical training or sports. Moreover, robotics devices have been used for the rehabilitation or re-education of the human's limbs [58, 59]. In this case, it is common that the task and trajectory are predefined and repetitive [60]. One of the challenges is to consider the variability between human subjects and, over time, to adapt the robot's behavior. Therefore, the challenge to be tackled is to consider the human model with time-varying parameters (such as joint impedance) and estimating them online to achieve predefined training objectives [61].
4. *Human-Robot sharing authority.* In this case, the human and robot should share the authority and thus switch the leadership when collaborating to accomplish a common task. In this case, the leadership may be taken alternatively by the human or the robot according to the status of the common task and a predefined collaboration protocol [62, 63]. This required that the robot have a certain level of autonomy (also called Level of Automation (LoA) [63]) while considering physical interaction with the human.
5. *Object/tool co-manipulation.* In this type of task, the human is considered as an exclusive leader, while co-manipulating an object or a tool with a robot. This may be seen as part of the previous situation when the human is the only leader. We consider here that the decision made by the human can not be made by the robot (otherwise the task may be done exclusively by robots). In this case, the robot should bring physical assistance to the human, by handling a large payload [64], reducing the human efforts [11, 65], avoiding non-ergonomic posture [66, 67] and by stabilizing the human movement [68]. Therefore, one of the main challenges here is how the robot brings this assistance by being active and by anticipating the human's intention, which is not predefined or known [69]. Also another challenge is to consider the co-manipulated objects/tools with a certain uncertainties of the parameters [32].
6. *Human-based learning.* In this situation, the robot is passive when the human is active, which permits the learning of human skills by the robot [23, 70, 71]. In this type of application, the human is commonly used as a "teacher" to train the robot, which becomes autonomous by *learning* how to playback the learned task.

In all these applications and contexts of pHRI, what can be defined as a pHRC is only a continuous physical interaction. Moreover, a collaboration, as presented here, involves a minimum mutual activity of each partner, which is not the case for all activities shown at the bottom of Fig. 2. Hence, pHRC may concern only situations 3, 4 and 5, where both actors (human and robot) are engaged in continuous contact for a period of time.

More details about pHRC that concerns the situation 5 (*Object/tool co-manipulation*) will be given in (Sect. 4), since it presents a important and actual challenges. Therefore, in next of current section, the continuous physical interaction is not presented, since it is addressed in the next section about a pHRC context.

As mentioned previously (Sect. 2), the classical robots were aimed to be separated from human operators by placing them in cages to avoid any possible risks of human injury. With the new definition of future industry (4.0 and 5.0) that allows a more flexible industry and human-centered paradigm (Sect. 1), one of the main improvements was the concept of workspace sharing and co-existence between humans and robots. With this new approach, the possibility of a robot's physical interactions with other objects or robots is extended to the interaction with humans. This constitutes a new paradigm and an important changes and challenges in future robotics, but it brings additional issues mainly related to human safety and its conditions. Moreover, other issues due to involving a human in the *interaction loop* (compared to static or dynamical objects) may appear since the human may have an unpredictable and time-varying behavior but also react differently to the robot's actions.

Therefore, this forced developers and researchers to focus first on how to share space between robots and humans, while ensuring human safety, no matter the situation or task to be performed. To this end, considering the human as a dynamic object or an external force has been a good starting point [13, 42]. In this case, the human is considered as an external part of the system and, thus, a part of the environment. Consequently, the use of modern and advanced sensors to get and exploit information about the environment (including humans) in order to adapt the robot behavior becomes essential for space sharing and physical human-robot interaction.

For the sharing workspace, many different contexts have been considered, even in a non-industrial environment, with a remote system to assist in a supermarket or store to identify and pick up requested elements [72, 73]. Considering the great difficulty of including the human in the system, previous works integrated cognitive aspects for interpreting and predicting its intention [74, 75].

In all these cases of sharing space, two main cases may be considered and presented next, human-robot collision avoidance and punctual contact detection for handling of the robot reaction.

3.1 Collision Avoidance

When sharing space between the human and the robot, collision avoidance seems to be an intuitive and less complex solution to ensure human safety. Therefore, in several collaborative robotics applications, immaterial separation is commonly used [76], which consists of the use of optical or visual sensors to detect a human's presence and thus adapt the velocity or stop the robot's movement. We can point out that this solution presents less flexibility and is close to classical robotics with materialized cages since it does not permit a parallel working of both humans and robots in the same shared space.

For more flexibility, it can be more interesting to ensure this collision avoidance while keeping each actor active and in a dynamic environment. However, this presents a more challenging approach. Indeed, there are difficulties in the prediction of human behavior, and even if its task is known, its behavior may present important uncertainties. Exploiting the human model can be an interesting approach. However, it also presents a high complexity, and thus, developing flexible strategies using this approach in different scenarios seems difficult to achieve. Moreover, considering the reactions of human and robot to the other partner's action, ensuring human safety still remain a challenge.

As an alternative, as mentioned before, considering the human as a dynamic object or an external force is an interesting classical and efficient approach to consider. Therefore, the robotic task, when sharing a workspace with the human, makes the environment unstructured. Hence, real-time acquisition and processing of the environment's information becomes fundamental for understanding the current state of the workspace.

One known solution consists of the use of vision system sensors (e.g. cameras) that provide rich information about the environment, which is one of the main features of modern robotics. For example, they are widely used for navigation in mobile robotics [77, 78], which integrated the collision avoidance aspects and potential field techniques [79, 80]. In a specific case of a manipulator robot, 3D depth cameras were commonly used to make the robot avoid possible collisions with humans to ensure its safety when sharing a workspace [7]. In that sense, several previous works exploited the Microsoft Kinect system and studied how to adapt the robot's behavior from its predefined trajectory to avoid collisions with dynamic obstacles (humans or objects) [7, 81, 82]. In these works, it is usual to retract the robot from the scene to detect and track only the real obstacles (by removing the noise in the image). Authors in [83] exploited the depth space to compute distances between the robot and dynamic obstacles in real-time and then, the robot was controlled using the potential filed principle by applying virtual repulsive forces. Thereafter, the obstacle-robot distance estimation methods were more deeply explored in [84] by improving and making the method faster for a real-time application.

However, using vision systems implies robot occlusion issues when the obstacle is between the robot and the camera. To overcome these issues, one approach might be to use multiple Kinetics to increase the workspace representation [81, 82]. Nevertheless, the use of multiple Kinetics increases the calibration complexity and the computational cost, which makes more complex the processing in real-time. To handle this issue, other approaches used one Kinect only by considering the obstacle with an infinite depth, called a *gray area* [7, 84]. This approach prioritizes human safety with a too-conservative behavior but highlights efficient results when the robot is not hidden by the obstacle. However, when the obstacle is placed between the robot and the camera, it will be considered close to the robot even when it is far from it in the depth axis, and thus it cannot deal with the case of obstacles that hide completely

Fig. 3 Setup of Human-Robot collision avoidance, where the robot may be hidden by the obstacle in the camera's view

the robot from the camera's view. Indeed, since the robot is perceived by only the camera, when it is hidden, its posture or end-effector cannot be estimated.

To overcome this issue, this problem was addressed in [8, 42] by exploring different approaches as an alternative to the use of the *gray area* approach, which is thus less conservative. An image illustrating the setup to study this issue can be seen in Fig. 3, where a robotic manipulator may be hidden (in the camera's view) by a moving obstacle.

In this case, the approach consisted of using not only the Kinect camera but also the robot's kinematic model and proprioceptive data (i.e. joint position measurements). The main scheme of this approach is summarized in Fig. 4, and more details and results may be found in [8, 42]. By combining both the vision system with the robot's proprioceptive sensors, the position of the robot with respect to the camera is known and updated at each instant, even when the robot is hidden from the camera view, and this perception does not suffer from any problem of noise, usual in image processing domain.

From that point, the collision avoidance strategy is improved using what is called a *safety contour* around the object, instead of considering an infinite depth (*gray area*) safety. Results showed that with a rough knowledge of the object's size (to define the *safety contour*), this method provides a good and smooth behavior of the robot to avoid collisions, even when it is hidden behind the object. However, in this considered work for instance, only the end-effector of robots is considered as a part to be avoided and the most dangerous in the interaction with humans. This is a limitation, compared to the previous known works [7], which needs to be improved.

Despite the fact that many collision avoidance strategies have been developed and experimentally tested with good results, there are still limitations to the success of collision avoidance scenarios. Indeed, if the velocity of the robot is limited by the safety regulations of cobotics, when the object (and human) moves faster than this limit, the collision strategies might fail. Moreover, in all these strategies, some

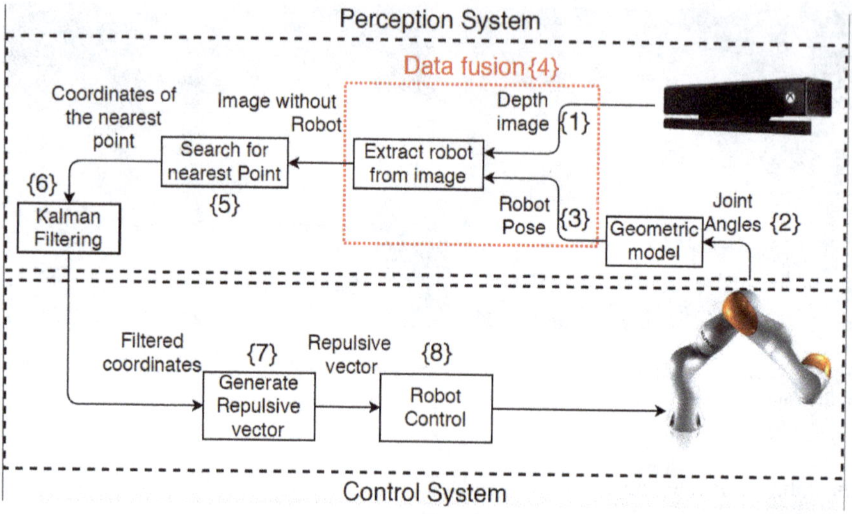

Fig. 4 Principle of H-R collision avoidance, based on fusion depth-camera and proprioceptive robot data [42]. The top part focuses on perception and distance estimation, whereas the bottom part focuses on the control strategy that ensures collision avoidance

parameters (such as ones related to the virtual repulsive forces) were set experimentally and based on a specific defined task, and it seems difficult to define a generic set of parameters, satisfying any kind of tasks and interaction behavior.

It is important to point out that vision-based systems are commonly used in a collaborative robotics context, not for collision avoidance only, but to perceive the environment in order to assist and monitor the collaboration. They were used to identify the task to be accomplished or the needed tool to be shared with humans. Hence, in [85], visual feedback is used to detect boxes and the localization of the pallet in a warehouse for moving and placing the object with a mobile arm manipulator. Also, the vision sensor is used to monitor a collaborative assembly process and detect the stage of the task as in [86].

3.2 *Collision Detection and Reaction*

For the purpose of co-existence between humans and robots through workspace sharing, collision avoidance may not always be ensured, and punctual physical contact may be inevitable. Therefore, it is always necessary to handle collisions, in case it happens. In that sense, a framework considering collision detection and reaction was developed in [13], where other aspects were considered as well, such as collision avoidance and possible collaboration. Handling collisions includes, here, detecting when the collision happens and then applying an appropriate reaction of the robot that avoids harming the human or, at least, reduces negative consequences. It also means ensuring appropriate behavior of the robot that makes physical contact with the human less dangerous through a compliant behavior of the robot (see Sect. 4).

The collision detection between a robot and any external part of the environment (including humans) has been deeply explored in [9, 13, 87] by exploiting fault detection and diagnosis techniques on the robot when detecting an abnormal contact [88]. This collision detection includes, in reality, not only the detection (when the contact happens) but also isolation (detecting where the contact happens on the robot) and identification (collision with the end-effector or with the robot body). Hence, a few techniques that avoid the use of additional sensors explored energy observer based on *generalized momentum* and the use of *residual vector* [88].

Under the condition of collision detection, the reaction strategy of the robot is explored as well. Different from the classical concept of robotics, where handling a collision means stopping the robot and the task in process, in the context of pHRI, its meaning is much deeper as described in [13, 89]. While working along with a human, the idea is to accept collisions as a possible part of the interactions, meaning that not only detecting them is important [89], but also the robot's reaction and behavior must be adapted in order to do not harm the human when a collision appears. Therefore, several strategies of reactions have been explored and experimentally tested. Hence, the robot may be stopped or moved back as a reflex along the direction of the collision vector [9, 87]. Also, the redundancy of the robot may be exploited to react while keeping its current task [90]. Meanwhile, a study about the effect on humans of those collision and reaction techniques has been done in [9, 91]. The authors provided a formal study of the quantitative effects of those impacts on the arm and chest of a human.

Adapting the robot's reaction and behavior may not be enough to ensure human safety, and considering a possible collision leads to making the robot's behavior compliant every time to permit a human-friendly interaction. This is especially the case when humans and robots interact continuously and collaborate. This compliant behavior is based on new design of robots and their actuator [26, 27, 92], but also on robotics advanced control techniques [32]. These compliant control strategies, including force control and impedance control [30, 93, 94], consider the forces exchanged with the environment while performing a task. By doing so, the robot will intend to perform the desired trajectory, but if a contact or external forces appear, the task will be somehow adapted, giving a certain degree of compliance to the robot's behavior.

In that sense, several previous works explored different situations and how to handle physical interaction between a robot and its environment. [95] proposed a combination of methods to generate a compliant control when preventing overshooting to manage the contact condition of the robot's end-effector with an unstructured environment (example of a polishing task). With a similar goal, the authors in [96] proposed an approach for handling the moment when passing from no-contact to contact with a surface and the possible instabilities that might arise for a manipulator.

Another approach to extend the compliant behavior in the presence of an unknown environment is to use learning methods, such as in [97], where repeating the task allows learning and adapting the control to have the desired impedance-like behavior despite the unknown environment. This approach can be useful if tasks are repetitive since the robot can learn from previous repetitions. Nevertheless, it might be a limiting method when the environment changes continuously, especially if a human is involved.

In the context of the robot's compliant control, having interactions with unknown objects or with a human can be seen as a similar issue since, in both cases, the task should be adapted to not damage the robot or object and not harm the human. Nevertheless, a particular difference can be pointed out as the human movements are unpredictable and might be fast, forcing the robot to depend more on its own behavior and less on the knowledge of the environment. Another difference is linked to the risk-level acceptability, which may be different for the robot/object and the human.

These compliant control techniques have been explored by adapting online the robot trajectory based on forces applied to the end-effector, as studied in [98, 99], in the case of the mobile manipulator.

As we can notice in previous works, the limit between punctual and continuous physical interaction may be very thin, and the transition between them should be made smooth, thanks to compliant control techniques, to allow a physical collaboration [13]. Therefore, the two situations (punctual and continuous interactions) and the way to handle them should be complementary since the continuous interaction implies a punctual one at the beginning. Moreover, the continuous physical interaction, as mentioned before, is usually related to the physical Human-Robot Collaboration (pHRC), which is presented in the next section.

4 Physical Human-Robot Collaboration (pHRC)

As mentioned previously (see Fig. 1), pHRC is a part of pHRI since pHRC implies a physical interaction. However, in pHRC, robots and Humans are sharing a task, which means they have to share the same purpose and accomplishment during the collaboration. This is particularly interesting where the contact is expected, targeted and should be handled continuously. There are already existing previous fields where the continuous physical collaboration between the human and the robot might have clear applications. As an example, we can find physical interaction through the teleoperation of the robot by the Human in order to control and have a force feedback of the robot, using a haptic device [1]. However, this case might not be considered as a pHRC as defined here, since even if teleoperation involves collaboration, with a very low level of robot leadership, the Human and the robot are not sharing the same workspace. Indeed, the current pHRC definition means a continuous physical interaction, when the human and the robot are sharing the same task but also the same space, such as in the classification presented in Sect. 3.

Moreover, one of the known robotics fields that might be considered as a pHRC concerns the active exoskeletons to enhance the human force capabilities [24, 100].

In this previously existing field, there is somehow a simplification in the robot control strategy, which is handled in the joint space as the robot fits the human body shape, while in other pHRC applications, human and robot are connected only in the task space (through each end-effector). Hence, the joint space trajectories and behaviors of the robot are not necessarily symmetrical regarding the human posture, where the robot's kinematic singularity still remains an issue to be tackled [101].

Therefore, with respect to the pHRI, pHRC is addressed separately in this section since it involves challenges that are not posed in other cases of pHRI. Indeed, the mutual activities of each partner to accomplish the shared task pose several problems and issues to tackle. The main challenge is related to the coordination strategy between humans and robots and how they share the leadership during the collaboration. This issue has been addressed in literature as a collaboration between humans and machines in general [102], and related to what is defined as a *level of automation* or *level of autonomy* (LoA) of the machine [63]. In that sense, the link with collaboration between robots (multiple robots) [103] is not relevant to make. Indeed, unlike multi-robots where a common system may coordinate the collaboration, in pHRC, the human is not controlled and thus coordination becomes more complex.

Other issues, that increase in pHRC, concern the interaction stability when the two partners (human and robot), not coordinated with a main common system, are both active and thus add actions on the whole system and may react to the action of the other. This last issue is directly linked to and may impact the human safety [12].

The deep definition of pHRC and its issues depends greatly on the collaboration task. Hence, in the next subsection, pHRC will be addressed regarding task-dependent collaboration to give an overview of problems and constraints related to the task type. In that sense, a key task in pHRC consists of object co-manipulation by humans and robots as detailed in Sect. 4.2. This type of task gathers several applications, such as for co-lifting of a load or co-manipulating a tool, which is an interesting field of the actual research. Finally, to evaluate the interest of pHRC, it is important to focus on its assessment, especially from the human feeling and point of view. Hence, it is crucial to evaluate the positive (or negative) impact of robots to justify the collaboration. Hence, some criteria and techniques for evaluating the collaboration's quality are addressed in Sect. 4.3.

4.1 Task-Dependent Collaboration

One of main questions when dealing with pHRC is what should be the role and collaboration level of the robot? Should the robot be a simple follower or a slave, as in teleoperation? Or could it take the leadership of the collaboration, and in which condition? If the robot can be the leader of the whole task, what becomes the role of the human? Could it be replaced by another follower robot? This field is known as *role assignment*, as addressed in a recent review [104], where some of the most meaningful topics (e.g., use of human-human collaboration to design the human-robot collaboration or how the shift in the role influence the collaboration) are discussed.

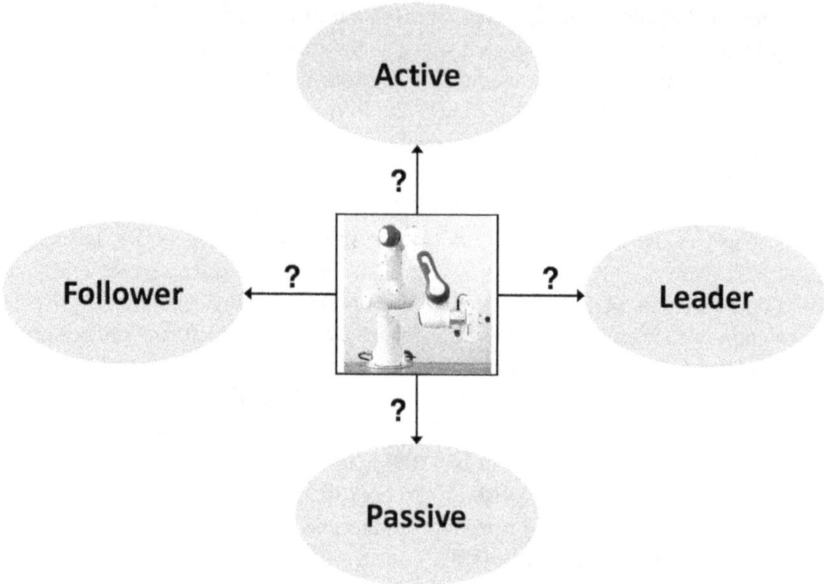

Fig. 5 Representation of the role assignment of robot during the physical collaboration with human [105]

Therefore, this role or level of assistance provided by the robot (as shown in the scheme of Fig. 5 [105]) is rather difficult and has been commonly defined case by case, depending on the task. In the context of pHRC, the role can be seen as how active the robot is during the collaboration, with respect to being a simple follower.

Depending on the task and thus on the role assignment, several objectives are researched in collaboration. One important aspect of pHRC concerns transparency, which is the capacity for a robot to follow human movements without any human-perceptible resistive forces [69]. When co-manipulating a heavy object, for example, it corresponds to the feeling for the human operator to not carry it out or just a small part, which should decrease the human's efforts.

In that sense, the authors in [106] studied the role assignment while co-manipulating a heavy table with an important size. In this work, the authors compared dynamic with fixed role assignment approaches and analyzed the perception of humans based on quantitative measurements. Hence, different role assignment strategies could also depend on the experience of the human, even for the same task. Besides, the change in the role assignment during the task might be an interesting point to explore. However, it might degrade the collaboration since it cannot be intuitive from the human's point of view, especially during the transition. In a similar context of manipulating large and bulky objects, the authors in [10] allowed the robot to choose from a set of possible assistance and thus somehow define the role assignment. Thus, they could define online, based on a library with previously developed methods, the best assistance based on the human's behavior. As a concrete application, a human-robot collaboration is used for an assembly task in the car industry by improving the human work condition [86].

Another aimed role for robots that may improve pHRC concerns pro-activity, which consists of the anticipation of the human intention and movement [107]. In that sense, the difference between reactive and proactive behavior is addressed in [108], where a reactive follower behavior is extended with online human movement predictions to allow a proactive behavior of the robot when co-manipulating an object with a human. Results show how a pro-active behavior of the robot based on an accurate estimation might help to improve collaborations for the human partner, which is one of the main purposes of pHRC. However, if this pro-active behavior is based on less accuracy than human knowledge, this might lead to worse results for the collaboration than a simple robot-follower (e.g., unnatural or uncomfortable behavior from the human's point of view). This issue may increase since humans do not behave similarly, according to their level of expertise on the task. This makes the switch between appropriate active-passive robot behavior very complex since it is not only task-dependent but also human-subject-dependent.

Therefore, several works about pHRC that aimed at a pro-active behavior of the robot focused on human intention and its detection. In the work of [109], the authors studied how the knowledge of the velocity profile in human movements can be used to enhance the robot's performance in physical collaboration by reducing considerably the human's effort, with a certain degree of assistance. Nevertheless, this study considered only one degree-of-freedom system in a point-to-point movement, which presents a quite controlled environment.

It seems justified that one of the main research objectives in setting robot behavior in pHRC is to define a *generic behavior* of robot that satisfies the physical collaboration with the human in most cases. However, independently from the type of robot, its targeted behavior and its role assignment depends necessarily and greatly on the type of the task and its objectives, as shown in those previous works.

For this reason, a varied set of predefined types of assistance are usually used, selected, and combined. Another example highlighting this strong dependency on the task is the work done in [110], where the authors achieved good results for collaborative sawing and polishing. In this work, the role assignment of the robot changes through the task as it starts as a *follower* to learn the movement patterns, and when the human starts showing signs of muscular fatigue, the role of the robot becomes more active. With this framework, the authors allow the robot to learn the task during its passive phase in order to become more active and to take the lead thereafter.

It can also be interesting to define a framework when the robot co-accomplishes with a human a shared task in a pHRC context but with the possibility to take the lead of other sub-tasks where the human contribution is not needed. The difficult aspect, in this case, concerns the switch between the two modes, where the physical collaboration requires a compliant behavior and tuning of some specific parameters to ensure safety, while the classical robotic behavior requires a "rigid" behavior (without compliance) to ensure an accurate behavior and a good trajectory tracking. To this purpose, a framework that handles a switch between role assignments according to the task phase was developed in a previous works [32], where the co-manipulation of objects with unknown dynamics was considered. In the next section, the specific aspect of pHRC to co-manipulate an object is addressed.

4.2 pHRC for Object Co-Manipulation

Among the many possible tasks that can be achieved through a pHRC framework, the co-handling or co-manipulation of an object jointly by the human and the robot seems to be an important and key task but with several additional complexities. Mainly, the robot may physically assist humans by reducing the carrying load [65] and also by stabilizing the movement when co-manipulating a tool [111].

As illustrated by Fig. 6, in the pHRC framework for object co-manipulation, we can distinguish different parts of the whole system: the robot, the object and the human. Considerations of each of these sub-parts of the pHRC framework may introduce challenges and need particular attention, related, for example, to the object particularities (its size, uncertainties, deformability,...). Integration of the human in simulation as a way to validate some approaches are also challenging, where the simulation of its behavior and its high variability presents an important issue.

In order to achieve this type of task, several additional challenges and issues, mainly introduced by the co-manipulated object, should be tackled, such as how to grasp the object, how to handle bulky or deformable objects [112, 113] and how to handle objects with important or unknown load [32, 64]. Other issues about the co-manipulated object concern representation singularities of its orientation in the context of compliant control [114]. Moreover, the object co-manipulation poses other challenges to human and robot coordination, which depend strongly on the task to be done and its objectives, as mentioned previously.

Since the object co-manipulation between humans and robots involves a high level of complexity, the motivation of using pHRC to accomplish a given task seems

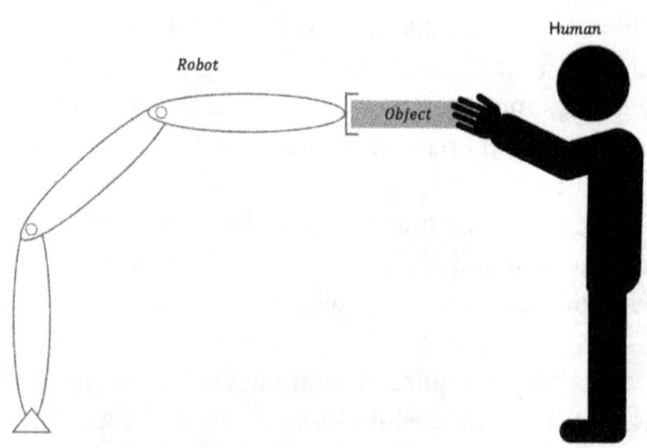

Fig. 6 Object co-manipulation in the context of pHRC (from [105])

relevant. Therefore, several existing works applied object co-manipulation under pHRC by giving particular attention to its benefits, especially from the human point of view. Hence, in [115], the human's exerted forces are evaluated in an object positioning collaborative task with the robot behavior improvement. In reality, to specify and get the benefits of using pHRC to co-manipulate an object, different issues mentioned above have been tackled in several previous works.

One of the main problems is to handle the payload introduced by the co-manipulated object. Hence, authors in [116] investigated how the payload is shared in a collaborative task of object co-manipulation when human is guiding the manipulator's end-effector in a simple trajectory. They consider the object weight as unknown, which is an interesting assumption and more realistic while doing an object co-manipulation. The important payload of the co-manipulated object has also been tested in [64] using a heavy 4-DOF mechanism. In this study, the limits of the typical controllers were investigated to avoid vibrations and dangerous behaviors in a pHRC context. Even if their device differs from a classical manipulator to transport an important payload, the study shows again the benefits of considering the payload as known in the controller design.

Other issues to tackle in every pHRC task concern the movement coordination between the human and the robot, since the synchronization of their movement is important for an efficient collaborative task. In a particular case of object co-manipulation, its size and complex geometry bring additional problems to the coordination, especially when no vision sensors are involved. Even if the weight and dynamics of the object may present some uncertainties, its size and geometry are usually supposed to be known. Therefore, to handle the coordination, human motion estimation was addressed in [117] by exploiting the minimum jerk model. In this study, an additional interesting aspect of the energy exchange between the human and the robot is analyzed, where *unnecessary energy* is referenced as the parameter used to improve the collaboration.

In the same sense, to improve this coordination, an anticipatory principle of human behavior was introduced and explored in previous works to adapt the robot behavior by using its kino-dynamic states in real-time [69, 118].

This anticipatory principle of human behavior may also be combined with knowledge of the task and its objective. As an example, the objective of keeping the object horizontal was targeted in [108] between a human and a humanoid robot. In this case, the robot should have the pro-activity capability, as mentioned in the previous section.

This human movement anticipation during the collaboration refers in literature to human intention and its detection. In that sense, distinguishing the human intention to perform translation or a rotation of an object is a problem tackled in [20] by developing a theoretical framework that controls and coordinates movement to precisely pose an object in a collaborative assembly task. This translation/rotation ambiguities in a collaborative object co-manipulation have also been addressed in [119].

These ambiguities are even accentuated if the object is long, and thus, the forces and torques are sensed at the robot's end-effector. This is due to difficulties in perceiving the force/torque of human interaction when the sensors are placed far away (on the robot side). To tackle these problems, authors in [120] classified the intention mode (translation and rotation) using the magnitude and direction of the force.

Hence, the human intention estimation seems very important, especially when co-manipulating a heavy and bulky object, as it was investigated in [121] to achieve a collaborative lifting task. In this work, the manipulator's control improvement has been introduced to better assist the human during physical collaboration. Their approach to identifying the human's intention is similar to [122], but by adapting the controller with the object information and improving the robot's behavior.

To investigate strategies for object co-manipulation using pHRC, inspiration for physical collaboration between humans seems obvious. Hence, works of [123] studied the way two people manipulate an object, to then use that information in [124] to adjust a robot's controller and to co-manipulate an object with a human. Finally, in [125], an optimal damping behavior is proposed for that task. Similarly, in [126], an experiment is conducted with two persons, where a protocol of master-slave is defined and respected. This allows us to obtain a behavioral pattern that will be transferred afterward to a follower robot. In this study, a vertical movement is applied, with the objective of keeping the object in a horizontal position. More recently, authors of [127] introduced a model derived from human arm movements theories to compute the interaction force during an object co-manipulation task, by two persons and by only one using both hands. This joint manipulation that involves coordination between two entities is often designated, as in this work, as *dyadic* collaborative manipulation.

In a practical and concrete application, object co-manipulation is also addressed by considering different steps and situations and by integrating multiple robots and humans. Therefore, in the works of [115], different states of the collaborative object manipulation task were considered, such as the starting, the moving, and the final positioning of the object. To consider situation changing, works of [121] introduced several adaptation methods according to different robot's goals. The authors proposed a framework for an assembly task, including environment detection, object grasping, and collaborative assembly. Moreover, a more general framework for multiple robots and humans manipulating a bulky object is proposed in [112]. This framework, which includes more than two agents, makes even more complex the coordination issue, as discussed in this work.

As mentioned in these previous works, the purpose of the collaboration and task is important to justify the robot's presence and to define the best robot's behavior. Hence, it is important to consider the object to manipulate or the goal of the collaborative task while designing the robot control. On the other hand, it seems convenient to have a generic robot's behavior adaptation that takes into consideration the different working modes and states in the collaborative process. Hence, a tradeoff is to find between the genericity of the robot's control method and its efficiency in terms of objectives, which is strongly task-dependent and thus requires more knowledge about the task.

In that sense, recent works considered the co-existence and combination of classical and repetitive robot behavior with continuous collaborative and compliant behavior [32, 105]. In this work, a simple framework was designed and consists of a manipulator robot capable of achieving repetitive tasks on its own in a classical way, where it is much less compliant (more rigid), as well as collaborating with the human (with more and adaptable compliance) whenever the latter desires to collaborate and without presenting any risk. This is very useful by making a co-existence of classical and collaborative robotics by adapting its behavior for the purpose of the same task. In this work, a human is moving, along with a robot, an object that presents an unknown payload in physical collaboration (pHRC), and thus it cannot be included in the model of the robot. Hence, the framework exploits a force sensor placed on the human side for force interaction estimation in order to help humans by reducing its feeling payload. The robot control and its behavior adaptation principles, particularly the compliance, are detailed in [32]. Some considerations related to the object unknown payload can be found in [128].

During a collaborative co-manipulation of objects, the goal is to keep humans behaving as much as possible in a natural way. One good collaboration may be the intuitive transparency, which consists, in the case of object co-manipulation, of the feeling of a human moving alone (without the object and without the robot). Therefore, it is important that a human performs a trajectory without applying important efforts, which gives him a natural feeling in the movement. In addition, avoiding uncomfortable postures for humans during long collaborations is an important purpose to achieve.

More generally, the justification of robotic-based collaboration requires an evaluation of its benefits through a clear and explainable metrics or assessment criteria. Hence, to evaluate and compare different scenarios and controllers for a "good" collaboration in general and particularly when co-manipulating an object, some metrics should be proposed based on information on the interaction forces and feedback from the human.

More details and discussion about these metrics are addressed in the next section about the collaborative task in general. However, in the case of an object co-manipulation task, the metrics depend on the object and on the task's objective. More details about metrics in the pHRC context can be found a previous work [11].

4.3 *Evaluation of Collaboration's Quality*

In the context of pHRC particularly, the *quality* of the collaboration is an important aspect to define and consider in order to justify its existence and advantage. In that sense, the concept of quality could include multiple aspects, where the human feeling and its point of view seem to be crucial.

Therefore, several quality criteria that define the "good" physical human-robot collaboration can be stated, such as the *human safety*, the *quality of assistance*, and the *performance* to accomplish the common task. Some of these quality criteria may be inspired by the criteria of physical human-robot interaction [14].

- The *human safety* consists in reducing as much as possible the risk for the human related to the continuous physical contact with the robot. In that sense, two complementary aspects are usually addressed. Firstly, the design of softer and compliant robotics is developed [26, 34], and the risk for the human is evaluated to define the current safety legal standards (Sect. 2). Secondly, the stability of interaction is considered one of the most important aspects that should be ensured in the robotics control techniques [129]. By using existing robots designed for physical collaboration, the robotic control techniques and collaboration strategy are the main research ways to investigate to ensure human safety in the pHRC context.
- The *quality of assistance* concerns how much the pHRC is useful for human with respect to working alone and bring a positive impact in a few aspects such as the level of assistance, physical and cognitive comfort, ergonomics improvement, and intuitiveness of interaction. Therefore, through these criteria, the collaboration should ease the task for the human or allow him to do it in a comfortable, better, and easy (intuitive) way.
- The *performance* criterion in pHRC, as mentioned here, concerns the quality of task accomplishment. It is thus depending strongly on the task to be performed. In that sense, several studies focused on understanding what kind of movements should be performed by the robot (or a robot-like system) while working with a human [109, 127]. With respect to the task's objectives and our knowledge of classical robotics, these performance criteria can include different aspects, such as a high success rate, task speed execution and trajectory tracking accuracy. We can expect that these performance criteria in pHRC are comparable with the one obtained with classical robotics. However, it might be difficult in reality to reach these classical robotic criteria with a human-robot collaboration, considering the human and root constraints in the pHRC context (speed limits, more uncertainties, compliance behavior,...).

Hence, it is, for example, pointless to replace two robots that handle well a task, by human and robot in collaboration to handle the same task. Nevertheless, when the physical presence of human in the loop is required for some applications, the pHRC is justified and inevitable. This highlights that pHRC is complementary to classical robotics and is not intended to replace it completely, even in future industries.

Even if safety is an unquestionable point in pHRC, a tradeoff has often to be find between different *collaboration quality* criteria mentioned above with respect to global task objectives.

Analyzing the collaboration's quality in pHRC consists actually in evaluating the improvement in the task, mainly from the human's point of view. Therefore, a common approach is to evaluate the pHRC framework based on the feedback from human subjects, as done in [122, 124, 130–132]. In general, these evaluations, done over a large number of human subjects, can be qualitative (based on a human's feeling) or quantitative based on objective elements (e.g., error in following a path, time to achieve a task, reduction of human's effort,...).

It is important to point out that the most common approach for evaluating the collaboration's quality is qualitative, based on a survey of feedback from the human subject involved in the experiments, to evaluate and compare different aspects of the collaboration [85, 121]. To have a good evaluation with significant data and statistically relevant, a large number of subjects is usually required, which makes experiments complex and time-consuming.

Although approaches based on qualitative criteria are the most used in the literature, they have some drawbacks, mainly related to the availability of a large number of subjects and ethical considerations. Another drawback of those approaches is related to the dependency of criteria and defined questions on the type of task, its environment and its context, which may decrease the consistency of evaluations and thus introduce a high subjectivity.

For those reasons, it becomes interesting to have some quantitative metrics that might be used to assess and evaluate the collaboration's quality in the pHRC context. This is a challenging issue since human perception and feeling are particularly complex to formalize and transform on quantifiable criteria. To that purpose, in some previous works [85, 122, 130], the time needed to perform a task is used to evaluate the collaboration. This metric can be useful to evaluate a clear and repetitive task with small variations. Nevertheless, in the case of pHRC context involving different subjects with high variability, the task duration does not seem to be the most important criterion to analyze. As an example, the task duration might not be the most important metric when co-manipulating objects. For instance, a task could be accomplished rapidly but requires too much effort from the human, and thus decreases the other criteria of collaboration quality.

Other quantitative metrics based on efforts applied by the human at the end-effector were considered in [133], while authors in [134] evaluated the overload due to the task on human joints. In this context, other previous works [11], proposed and analyzed improvements in pHRC context using different quantitative metrics, such as the total force applied, the total energy injected by the human, and the positions and velocities of the robot. In this study applied for human-robot object co-manipulation, two cases when the payload is included or not in the controller were analyzed and compared based on the following quantitative metrics:

- The root mean square (RMS) of the wrenches applied by the human during the collaboration.
- The root mean square (RMS) of the power injected by the human during the movement, as well as the percentage of that power that has a negative sign.
- The root means square (RMS) of the tracking errors with respect to the desired reference trajectory.

More details about this assessment study may be found in [11, 105]. The first metric considers the total forces applied by the human; this could be understood as considering the integral of the forces applied by the human throughout the whole collaboration. Then, the goal is that the human performs the task with the most minor possible forces applied. However, from a practical point of view, comparing the area below the force curve of different experiments is not straightforward since each experiment might have a different duration or sampling time. Therefore, the mean value of the area below the curve would be more appealing. Therefore, these existing assessment tools should serve to evaluate the quality of any pHRC method, and in a way to evaluate the robotics proposed strategy of physical collaboration with human [32].

All these types of qualitative metrics should represent objective criteria. However, the most interesting are those consistent with qualitative metrics based on human feeling and in some defined context. Indeed, although these last metrics have some drawbacks, as mentioned above, they have an essential advantage regarding acceptability and ethics since, in human-involved experiments, criteria based on human feelings are expected to be trusted and accepted.

5 Human-Centered Robotics

In collaborative robotics and particularly in the context of pHRC, the human condition and feeling is of high importance, as mentioned above. Therefore, it is common to consider this robotics field as centered on human [135], and thus as a part of the human-centered robotics domain well known in literature [136].

In this case, the collaboration aspects, such as purpose and benefits, risks and limits, should be considered particularly from the human point of view. Under these considerations, different aspects of human-centered robotics with their issues and challenges can be found in pHRC applications. Hence, two main aspects that require a great attention are:

- How does the collaboration impact the human for doing a given task in terms of role assignment, benefits, limits and acceptability?
- How should the Robot behave when considering its collaboration with a human?

5.1 What Impact Has the Collaboration on the Human?

To consider the collaboration impact from the human point of view, the perception of this collaboration and its feeling is very important even if it presents some subjectivity, as mentioned previously (Sect. 4.3). From this point of view, the objectives of this collaboration should be clearly addressed. Hence, is the robot really needed to help in accomplishing the task? Inversely, if the robot may accomplish a given task, is the human needed to collaborate with the robot when it necessarily reduces its dynamic and its productivity as a classical robot? Therefore, the question to be asked when addressing the human-robot collaboration is what the robot may bring to the human for a collaborative task, and *vice-versa.*

This impact on humans depends also on its role in the collaboration. This role is to be defined or adapted with respect to the human accomplishing the task alone. This is related to the leadership sharing strategy and who handles the timing of collaborative tasks. Indeed, defining the human role can not be done without considering certain ethics and acceptability aspects since it seems difficult to consider a robot as a high-level leader and a human as a simple follower. Even if the robot may take a leadership sometimes during a collaborative task, this should always stay a *low level* leadership where the human may follow or resist any time to the robot as a supervisor and keeping thus its natural *high level* leadership. Also, for a better collaboration, it seems important that the human or robot understand and perceive the partner's intention. In this context, when for the robot, some methods have been investigated and developed to get the human intention [137, 138], the human also requires some learning to adapt to the collaboration. Indeed, the human reaction may ease the collaboration or make it more complex. This reaction may include several aspects, such as adaptation and learning aspect, inter-subject and intra-subject variability and human anticipation of the task (known or unknown). It seems clear that a human with more experience in collaboration with a given robot should make a more fluent interaction and thus a better collaboration [139, 140].

5.2 *Robot's Compliant Behavior*

Another consideration when humans have to collaborate physically with robots is how the robot should behave to seem *friendly* from the human point of view. Even if this aspect of human-friendly behavior, defined sometimes as human-like behavior, can have some subjectivity, there are some points that should be objectively guaranteed to ensure the safety and comfort of humans during the collaboration.

Therefore, to reduce the risk and dangerousness that robots may bring to humans, several efforts, research works and new developments have been done in recent years to change the robot in terms of design: lighter body, with mechanical compliance at joint level [26, 92] and even with flexibility at the body level [141], which is commonly called soft robotics. In order to complete this design evolution and adapt this compliance by making a good trade-off between the robot's efficiency and safety, the compliance aspects may be done through control techniques, mainly to make the robot behave as a mass-spring-damper system, which parameters may be varied through the controller.

The main techniques applied in robotics are impedance control and admittance control [64, 93]. A more concrete application of these techniques on pHRC and in the industrial-like context was addressed in [32].

For the same purpose of *human-friendly* behavior, one step to ensure safety consists in ensuring the stability of the interaction. One of the possible techniques that may be applied is passivity-based control. However, there is an important complexity due to the fact that the human may inject an *energy* which is not controlled or expected during the interaction.

In that sense, the classical passivity-based approach may be too conservative [14] since it requires passivity at each instant. Therefore, to improve the behavior of the system and maximize the capabilities of the robot, an alternative is about handling more efficiently the energy of the system. An interesting approach consists using the *energy tanks* formalism to allow certain non-passive operations but ensuring the passivity of the whole system in a certain *time window* [53, 142, 143]. This principle consists of including additional dynamics as energy reservoirs (called tanks) to preserve the passivity overall, even when some operations might not be passive.

The *energy tank* principle is addressed in several types of systems and increasingly in collaborative robotics and under a compliant-type control [131]. In that sense, and to address the energy aspects, the *Lagrangian-based* modeling is no longer adapted, whereas a *Hamiltonian-based* formulation seems more appropriate to express energy exchange. This have been combined in [144] with the *energy tank* formalism by designing a Cartesian impedance control in a closed loop when assuming flexibility of joint in the robot.

It is important to mention that human-friendly behavior, through compliance, has a price in terms of precision when accomplishing a task. Indeed, we can not imagine compliance and, thus, the flexibility of the robot without a *relaxation* of trajectory tracking. Therefore, the question is always about what is the trade-off between the compliance given to the robot, and thus the level of human-friendly feeling, and the level of robot precision.

From that point of view, it seems that collaborative robotics is not made to replace classical robotics, which offers more precision and high weight and dynamics in some classically automatized and repetitive tasks. It is intended to be complementary and applied to different types of tasks where human presence is needed for ethical reasons or to handle the complexity of perception, and the robot may bring motion stability and reduce human fatigue.

6 Conclusions

In the current chapter, we addressed the positioning of collaborative robotics (cobotics) in the whole context of robotics. We discussed thus how a given application can be considered as collaborative robotics and is not a system defined as a cobot, which makes a collaborative robotics application. Moreover, to make this robotics more collaborative, it is crucial to make it human-friendly by modifying the design of a new generation of robots and improving the control technique to make the robot more compliant, as addressed in several previous works in the field of collaborative robotics.

In this context, we defined the physical Human-Robot Interaction (defined in literature as pHRI) and focused on a more specific case of physical Human-Robot Collaboration (pHRC), which is not defined explicitly in the literature, where humans and robots share physically a common task. In this case, the physical interaction poses a certain number of problems related to human safety, sharing the leadership, ethics and the adaptation of the behavior of each partner regarding the other partner.

In the context of pHRC, we addressed the importance of the task and context in the way of dealing with the collaboration, the specific pHRC case of the co-manipulation object, and the importance of pHRC assessment.

In this chapter, we also discussed how the pHRI and pHRC will not be the future robotics, in competition with or taking place of the classical robotics, but as the future of collaborative and human-centered robotics. Whereas the classical robotics should remain more appropriate for several repetitive tasks and in a structured environment. Indeed, the robot are known to be accurate and fast, but with their classical use, these characteristics can not be ensured when collaborating physically with humans. This possibility of interaction has a "cost", since the robot in physical collaboration can not ensure the robotic performance as before but be more human-like or human-friendly.

Moreover, cobotics can not be expected to be a high productivity since it should be limited by the human rhythm. Therefore, a trade-off should always be found between the classical functioning of the robot (rigid and fast behavior) and compliant behavior with less accuracy and more flexibility.

As mentioned from the beginning, the human-robot collaboration addressed here is based, for instance, on only physical interaction and collaboration, which still presents an important challenges. However, there is another existing aspects or human-robot interaction such as social or cognitive interaction.

As a perspective of a more general, complete and better collaboration between human and robot, the integration and combination of all these communication modalities (physical, social and cognitive) should improve the future human-robot collaboration.

References

1. Hokayem PF, Spong MW (2006) Bilateral teleoperation: an historical survey. Automatica 42(12):2035–2057
2. Carrière G, Benoussaad M, Wagner V, Dessein G, Boniface B (2020) Off-line correction method suitable for a machining robot application to composite materials. Int J Adv Manufact Techno 110(9):2361–2375
3. Martínez-Díaz M, Soriguera F (2018) Autonomous vehicles: theoretical and practical challenges. Transp Res Proc 33:275–282. XIII conference on transport engineering, CIT2018
4. Ustundag A, Cevikcan E (2017) Industry 4.0: managing the digital transformation. Springer
5. Müller J (2020) Enabling technologies for industry 5.0 : results of a workshop with Europe's technology leaders (Commission European and and Innovation Directorate-General for Research), tech. rep
6. Gao Z, Wanyama T, Singh I, Gadhrri A, Schmidt R (2020) From industry 4.0 to robotics 4.0 - a conceptual framework for collaborative and intelligent robotic systems. Proc Manuf 46:591–599
7. Flacco F, Kröger T, De Luca A, Khatib O (2012) A depth space approach to human-robot collision avoidance. In: Proceedings - IEEE international conference on robotics and automation, pp 338–345
8. Nascimento H, Mujica M, Benoussaad M (2020) Collision avoidance in human-robot interaction using kinetic vision system combined with robot's model and data. In: 2020 IEEE/RSJ international conference on intelligent robots and systems (IROS), pp 10293–10298

9. Haddadin S, Albu-Schaffer A, De Luca A, Hirzinger G (2008) Collision detection and reaction: a contribution to safe physical human-robot interaction. In: 2008 IEEE/RSJ international conference on intelligent robots and systems, pp 3356–3363
10. Dumora J, Geffard F, Bidard C, Aspragathos NA, Fraisse P (2013) Robot assistance selection for large object manipulation with a human. In: 2013 IEEE international conference on systems, man, and cybernetics, pp 1828–1833
11. Mujica M, Benoussaad M, Fourquet J-Y (2020) Evaluation of human-robot object co-manipulation under robot impedance control. In: 2020 international conference on robotics and automation (ICRA). IEEE
12. De Santis A, Siciliano B, Luca AD, Bicchi A (2008) An atlas of physical human - robot interaction. Mech Mach Theory 43:253–270
13. De Luca A, Flacco F (2012) Integrated control for pHRI: collision avoidance, detection, reaction and collaboration. In: Proceedings of the IEEE RAS and EMBS international conference on biomedical robotics and biomechatronics, pp 288–295
14. Ajoudani A, Zanchettin AM, Ivaldi S, Albu-Schäffer A, Kosuge K, Khatib O (2018) Progress and prospects of the human-robot collaboration. Auton Robot 42(5):957–975
15. Khalil W, Dombre E (2004) Modeling, identification and control of robots. Appl Mech Rev 56:483
16. Colgate J, Wannasuphoprasit W, Peshkin M (1996) Cobots: robots for collaboration with human operators. In: Proceedings of the 1996 ASME international mechanical engineering congress and exposition, pp 433–439; Conference date: 17-11-1996 Through 22-11-1996
17. Colgate J, Peshkin M (1999) Patent no. 5952796. United States of America
18. Liu H, Wang L (2020) Remote human-robot collaboration: a cyber-physical system application for hazard manufacturing environment. J Manuf Syst 54:24–34
19. Azizian M, Liu M, Khalaji I, Sorger J, Oh D, Daimios S (2020) 3 - the da vinci surgical system. In: Abedin-Nasab MH (ed) Handbook of robotic and image-guided surgery. Elsevier, pp 39–55
20. Wojtara T, Uchihara M, Murayama H, Shimoda S, Sakai S, Fujimoto H, Kimura H (2009) Human-robot collaboration in precise positioning of a three-dimensional object. Automatica 45:333–342
21. Peternel L, Tsagarakis N, Ajoudani A (2016) Towards multi-modal intention interfaces for human-robot co-manipulation. In: 2016 IEEE/RSJ international conference on intelligent robots and systems (IROS), pp 2663–2669
22. Lauretti C, Cordella F, Guglielmelli E, Zollo L (2017) Learning by demonstration for planning activities of daily living in rehabilitation and assistive robotics. IEEE Robot Autom Lett 2(3):1375–1382
23. Ikemoto S, Amor HB, Minato T, Jung B, Ishiguro H (2012) Physical human-robot interaction: mutual learning and adaptation. IEEE Robot Autom Mag 19(4):24–35
24. Lee HD, Lee BK, Kim WS, Han JS, Shin KS, Han CS (2014) Human-robot cooperation control based on a dynamic model of an upper limb exoskeleton for human power amplification. Mechatronics 24(2):168–176
25. Navarro B, Fonte A, Fraisse P, Poisson G, Cherubini A (2018) In pursuit of safety: an open-source library for physical human-robot interaction. IEEE Robot Autom Mag 25(2):39–50
26. Albu-Schäffer A, Haddadin S, Ott C, Stemmer A, Wimböck T, Hirzinger G (2007) The DLR lightweight robot: design and control concepts for robots in human environments. Ind Robot 34(5):376–385
27. Vanderborght B, Albu-Schaeffer A, Bicchi A, Burdet E, Caldwell DG, Carloni R, Catalano M, Eiberger O, Friedl W, Ganesh G, Garabini M, Grebenstein M, Grioli G, Haddadin S, Hoppner H, Jafari A, Laffranchi M, Lefeber D, Petit F, Stramigioli S, Tsagarakis N, Van Damme M, Van Ham R, Visser LC, Wolf S (2013) Variable impedance actuators: a review. Robot Auton Syst 61(12):1601–1614
28. Migliore SA, Brown EA, DeWeerth SP (2005) Biologically inspired joint stiffness control. In: Proceedings of the 2005 IEEE international conference on robotics and automation, 2005. ICRA 2005, pp 4508–4513

29. Chu R, Zhang Y, Zhang H, Xu W, Ryu J-H, Wang D (2020) Co-actuation: a method for achieving high stiffness and low inertia for haptic devices. IEEE Trans Haptics 13(2):312–324
30. Hogan N (1985) Impedance control - an approach to manipulation. i - theory. II - implementation. III - applications. ASME Trans J Dyn Syst Meas Control B 107:1–24
31. Villani L, De Schutter J (2008) Force control. Springer, Berlin, Heidelberg, pp 161–185
32. Mujica M, Crespo M, Benoussaad M, Junco S, Fourquet J-Y (2023) Robust variable admittance control for human-robot co-manipulation of objects with unknown load. Robot Comput-Integr Manuf 79:102408
33. Ardiani F (2023) Contribution to the parametric identification of dynamic models: application to collaborative robotics. Theses, Toulouse - ISAE-SUPAERO
34. Bischoff R, Kurth J, Schreiber G, Koeppe R, Albu-Schäffer A, Beyer A, Eiberger O, Haddadin S, Stemmer A, Grunwald A Others (2010) The KUKA-DLR lightweight robot arm-a new reference platform for robotics research and manufacturing. Robotics (ISR). In: 2010 41st international symposium on and 2010 6th German conference on robotics (ROBOTIK), pp 1–8
35. Yang C, Ganesh G, Haddadin S, Parusel S, Albu-Schäeffer A, Burdet E (2011) Human-like adaptation of force and impedance in stable and unstable interactions. IEEE Trans Robot 27(5):918–930
36. Lumelsky VJ (1987) Dynamic path planning for a planar articulated robot arm moving amidst unknown obstacles. Automatica 23(5):551–570
37. ISO 15066 (2016) Robots and robotic devices- Collaborative robots. standard, International Organization for Standardization
38. Chemweno P, Pintelon L, Decre W (2020) Orienting safety assurance with outcomes of hazard analysis and risk assessment: a review of the ISO 15066 standard for collaborative robot systems. Saf Sci 129:104832
39. Marvel JA, Norcross R (2017) Implementing speed and separation monitoring in collaborative robot workcells. Robot Comput-Integr Manuf 44:144–155
40. Benhabib N, Padois V, Daney D (2020) Securing industrial operators with collaborative robots: simulation and experimental validation for a carpentry task. In: In Proceedings of the IEEE international conference on robotics and automation, (Paris, France). Finalist of the conference Best Paper Award in the Automation category
41. Wang L, Schmidt B, Nee AY (2013) Vision-guided active collision avoidance for human-robot collaborations. Manuf Lett 1(1):5–8
42. Nascimento H, Mujica M, Benoussaad M (2021) Collision avoidance interaction between human and a hidden robot based on kinect and robot data fusion. IEEE Robot Autom Lett 6:88–94
43. Bélanger-Barrette M (2016) Collaborative robots - risk assesment, an introduction. Robotiq, pp 1–20
44. Dumora J, Geffard F, Bidard C, Brouillet T, Fraisse P (2012) Experimental study on haptic communication of a human in a shared human-robot collaborative task. In: 2012 IEEE/RSJ international conference on intelligent robots and systems, pp 5137–5144
45. Francesco P, Paolo CG (2017) Aura: an example of collaborative robot for automotive and general industry applications. Procedia Manufacturing, vol 11, pp 338–345. 27th international conference on flexible automation and intelligent manufacturing, FAIM2017, 27-30 June 2017, Modena, Italy
46. Onnasch L, Roesler E (2021) A taxonomy to structure and analyze human-robot interaction. Int J Soc Robot 13(4):833–849
47. Brahmi B, Saad M, Luna CO, Archambault PS, Rahman MH (2018) Passive and active rehabilitation control of human upper-limb exoskeleton robot with dynamic uncertainties. Robotica 36(11):1757–1779
48. Mekki M, Delgado AD, Fry A, Putrino D, Huang V (2018) Robotic rehabilitation and spinal cord injury: a narrative review. Neurotherapeutics 15:604–617
49. Benoussaad M, Rakotondrabe M (2020) Feedforward and h feedback robotic force control in a 1-dof physical interaction using a nonlinear human model. IFAC-PapersOnLine 53(2):8531–8537. 21st IFAC World Congress

50. Yan T, Cempini M, Oddo CM, Vitiello N (2015) Review of assistive strategies in powered lower-limb orthoses and exoskeletons. Robot Auton Syst 64:120–136
51. Pons JL (2010) Rehabilitation exoskeletal robotics. IEEE Eng Med Biol Mag 29:57–63
52. Bryant A, Wei B, Veronesi G, Cerfolio R (2018) 28 - robotic surgery: techniques and results for resection of lung cancer. In: Pass HI, Ball D, Scagliotti GV (eds) IASLC thoracic oncology (Second Edition), 2nd edn. Elsevier, Philadelphia, pp 283-288.e1
53. Ferraguti F, Preda N, Manurung A, Bonfe M, Lambercy O, Gassert R, Muradore R, Fiorini P, Secchi C (2015) An energy tank-based interactive control architecture for autonomous and teleoperated robotic surgery. IEEE Trans Rob 31(5):1073–1088
54. Alouane MA, Huo W, Rifai H, Amirat Y, Mohammed S (2019) Hybrid fes-exoskeleton controller to assist sit-to-stand movement. IFAC-PapersOnLine 51(34):296–301. 2nd IFAC conference on cyber-physical and human systems CPHS 2018
55. Del-Ama AJ, Koutsou AD, Moreno JC, De-Los-Reyes A, Gil-Agudo Á, Pons JLA (2012) Review of hybrid exoskeletons to restore gait following spinal cord injury. J Rehabil Res Dev 49(4):497–514
56. Benoussaad M, Poignet P, Hayashibe M, Azevedo-Coste C, Fattal C, Guiraud D (2013) Experimental parameter identification of a multi-scale musculoskeletal model controlled by electrical stimulation: application to patients with spinal cord injury. Med Biol Eng Comput 51:617–31
57. Benoussaad M, Rotella F, Chaibi I (2020) Flatness of musculoskeletal systems under functional electrical stimulation. Med Biol Eng Comput 58(5):1113–1126
58. Vallery H, van Asseldonk EHF, Buss M, van der Kooij H (2009) Reference trajectory generation for rehabilitation robots: complementary limb motion estimation. IEEE Trans Neural Syst Rehabil Eng: A Publ IEEE Eng Med Biol Soc 17:23–30
59. Qian Z, Bi Z (2015) Recent development of rehabilitation robots. Adv Mech Eng 7(2):563062
60. Masiero S, Carraro E, Ferraro C, Gallina P, Rossi A, Rosati G (2009) Upper limb rehabilitation robotics after stroke: a perspective from the university of Padua, Italy. J Rehabil Med 41:981–985
61. Esposito G (2020) Human-robot physical interaction for adaptive sport activity
62. Draper JV (1995) Teleoperators for advanced manufacturing: applications and human factors challenges. Int J Hum Factors Manuf 5(1):53–85
63. Habib L (2019) Niveaux d'automatisation adaptables pour une coopération homme-robots. Phd thesis, Thèse de doctorat, Université Polytechnique des Hauts-de-France
64. Campeau-Lecours A, Otis MJ, Gosselin C (2016) Modeling of physical human-robot interaction: admittance controllers applied to intelligent assist devices with large payload. Int J Adv Rob Syst 13(5):1–12
65. Kim W, Lee J, Peternel L, Tsagarakis N, Ajoudani A (2018) Anticipatory robot assistance for the prevention of human static joint overloading in human-robot collaboration. IEEE Robot Autom Lett 3(1):68–75
66. Peternel L, Kim W, Babic J, Ajoudani A (2017) Towards ergonomie control of human-robot co-manipulation and handover. In: IEEE-RAS international conference on humanoid robots, pp 55–60
67. Busch B, Maeda G, Mollard Y, Demangeat M, Lopes M (2017) Postural optimization for an ergonomic human-robot interaction. In: 2017 IEEE/RSJ international conference on intelligent robots and systems (IROS), pp 2778–2785
68. Ueda J, Gallagher W, Moualeu A, Shinohara M, Feigh K (2017) Chapter Eleven - Adaptive human-robot physical interaction for robot coworkers. Elsevier Inc
69. Jarrasse N, Paik J, Pasqui V, Morel G (2008) How can human motion prediction increase transparency?. In: 2008 IEEE international conference on robotics and automation, pp 2134–2139
70. Wu Y, Zhao F, Tao T, Ajoudani A (Jan2021) A framework for autonomous impedance regulation of robots based on imitation learning and optimal control. IEEE Robot Autom Lett 6:127–134

71. Kronander K, Billard A (2012) Online learning of varying stiffness through physical human-robot interaction. In: 2012 IEEE international conference on robotics and automation, pp 1842–1849
72. Tomizawa T, Ohya A et al (2006) Remote shopping robot system,-development of a hand mechanism for grasping fresh foods in a supermarket. In: 2006 IEEE/RSJ international conference on intelligent robots and systems. IEEE, pp 4953–4958
73. Tomizawa T, Ohba K, Ohya A et al (2007) Remote food shopping robot system in a supermarket-realization of the shopping task from remote places. In: 2007 international conference on mechatronics and automation. IEEE, pp 1771–1776
74. Sakita K, Ogawara K, Murakami S, Kawamura K, Ikeuchi K (2004) Flexible cooperation between human and robot by interpreting human intention from gaze information. In: 2004 IEEE/RSJ international conference on intelligent robots and systems (IROS)(IEEE Cat. No. 04CH37566), vol 1. IEEE, pp 846–851
75. Roncone A, Mangin O, Scassellati B (2017) Transparent role assignment and task allocation in human robot collaboration. In: 2017 IEEE international conference on robotics and automation (ICRA). IEEE, pp 1014–1021
76. Karagiannis P, Kousi N, Michalos G, Dimoulas K, Mparis K, Dimosthenopoulos D, Tokçalar Ö, Guasch T, Gerio GP, Makris SP (2022) Adaptive speed and separation monitoring based on switching of safety zones for effective human robot collaboration. Robot Comput-Integr Manuf 77:102361
77. Pire T, Mujica M, Civera J, Kofman E (2019) The rosario dataset: multisensor data for localization and mapping in agricultural environments. Int J Robot Res 38(6):633–641
78. Kim J, Do Y (2012) Moving obstacle avoidance of a mobile robot using a single camera. Proc Eng 41:911–916. International symposium on robotics and intelligent sensors (IRIS 2012)
79. DeSouza GN, Kak AC (2002) Vision for mobile robot navigation: a survey. IEEE Trans Pattern Anal Mach Intell 24(2):237–267
80. Khatib O (1985) Real-time obstacle avoidance for manipulators and mobile robots. In: Proceedings. 1985 IEEE international conference on robotics and automation, vol 2, pp 500–505
81. Schmidt B, Wang L (2014) Depth camera based collision avoidance via active robot control. J Manuf Syst 33(4):711–718
82. Chen JH, Song KT (2018) Collision-free motion planning for human-robot collaborative safety under cartesian constraint. In: 2018 IEEE international conference on robotics and automation (ICRA), pp 1–7
83. Flacco F, Kröger T, De Luca A, Khatib O (2012) A depth space approach to human-robot collision avoidance. In: Proceedings - IEEE international conference on robotics and automation, pp 338–345
84. Flacco F, Kroeger T, De Luca A, Khatib O (2015) A depth space approach for evaluating distance to objects: with application to human-robot collision avoidance. J Intell Robot Syst: Theory and Appl 80:7–22
85. Lamon E, Leonori M, Kim W, Ajoudani A (2020) Towards an intelligent collaborative robotic system for mixed case palletizing. In: 2020 IEEE international conference on robotics and automation (ICRA). IEEE, pp 9128–9134
86. Cherubini A, Passama R, Crosnier A, Lasnier A, Fraisse P (2016) Collaborative manufacturing with physical human-robot interaction. Robot Comput-Integr Manuf 40:1–13
87. De Luca A, Albu-Schaffer A, Haddadin S, Hirzinger G (2006) Collision detection and safe reaction with the DLR-III lightweight manipulator arm. In: 2006 IEEE/RSJ international conference on intelligent robots and systems, pp 1623–1630
88. DeLuca A, Mattone R (2005) Sensorless robot collision detection and hybrid force/motion control. In: IEEE international conference on robotics and automation (ICRA), Barcelona, Spain, pp 1011–1016
89. Haddadin S, De Luca A, Albu-Schäffer A (2017) Robot collisions: a survey on detection, isolation, and identification. IEEE Trans Rob 33(6):1292–1312

90. Magrini E, De Luca A (2017), Human-robot coexistence and contact handling with redundant robots. In: 2017 IEEE/RSJ international conference on intelligent robots and systems (IROS), pp 4611–4617
91. Haddadin S, Albu-Schäffer A, Hirzinger G (2009) Requirements for safe robots: measurements, analysis and new insights. Int J Robot Res 28(11–12):1507–1527
92. Albu-Schäffer A, Eiberger O, Fuchs M, Grebenstein M, Haddadin S, Ott C, Stemmer A, Wimböck T, Wolf S, Borst C, Hirzinger G (2011) Soft robotics: from torque feedback controlled lightweight robots to intrinsically compliant systems. Robot Res 70:185–207
93. Hogan N (1984) Impedance control: an approach to manipulation. IEEE Am Control Conf 1984:304–313
94. Hogan N, Buerger SP (2004) Impedance and interaction control. In: Robotics and automation handbook. CRC Press, pp 375–398
95. Roveda L, Iannacci N, Vicentini F, Pedrocchi N, Braghin F, Tosatti LM (2015) Optimal impedance force-tracking control design with impact formulation for interaction tasks. IEEE Robot Autom Lett 1(1):130–136
96. Salehian SSM, Billard A (2018) A dynamical-system-based approach for controlling robotic manipulators during noncontact/contact transitions. IEEE Robot Autom Lett 3(4):2738–2745
97. Li Y, Yang C, Ge SS (2010) Learning compliance control of robot manipulators in contact with the unknown environment. In: 2010 IEEE conference on automation science and engineering (CASE). IEEE, pp 644–649
98. Losey DP, O'Malley MK (2017) Trajectory deformations from physical human-robot interaction. IEEE Trans Rob 34(1):126–138
99. De Santis A, Pierro P, Siciliano B (2006) The multiple virtual end-effectors approach for human-robot interaction. In: Advances in robot kinematics. Springer, pp 133–144
100. Lee H, Kim W, Han J, Han C (2012) The technical trend of the exoskeleton robot system for human power assistance. Int J Precis Eng Manuf 13(8):1491–1497
101. Carmichael MG, Liu D, Waldron KJ (2017) A framework for singularity-robust manipulator control during physical human-robot interaction. Int J Rob Res 36:861–876
102. Sheridan TB, Verplanck WL (1978) Human and computer control of undersea teleoperators. tech. rep., MIT Man-Machine Laboratory, Cambridge, MA
103. Schillinger P, García S, Makris A, Roditakis K, Logothetis M, Alevizos K, Ren W, Tajvar P, Pelliccione P, Argyros A, Kyriakopoulos KJ, Dimarogonas DV (2021) Adaptive heterogeneous multi-robot collaboration from formal task specifications. Robot Auton Syst 145:103866
104. Jarrasse N, Sanguineti V, Burdet E (2014) Slaves no longer: review on role assignment for human-robot joint motor action. Adapt Behav 22(1):70–82
105. Mujica M (2021) Physical human-robot collaboration for object co-manipulation based on adaptable compliant control. Theses, Institut National Polytechnique de Toulouse - INPT
106. Mörtl A, Lawitzky M, Kucukyilmaz A, Sezgin M, Basdogan C, Hirche S (2012) The role of roles: physical cooperation between humans and robots. Int J Robot Res 31(13):1656–1674
107. Grosinger J (2016) Proactivity in robots. In: Proceedings of the twenty-fifth international joint conference on artificial intelligence, IJCAI'16. AAAI Press, pp 3994–3995
108. Thobbi A, Gu Y, Sheng W (2011) Using human motion estimation for human-robot cooperative manipulation. In: 2011 IEEE/RSJ international conference on intelligent robots and systems. IEEE, pp 2873–2878
109. Corteville B, Aertbeliën E, Bruyninckx H, De Schutter J, Van Brussel H (2007) Human-inspired robot assistant for fast point-to-point movements. In: Proceedings 2007 IEEE international conference on robotics and automation, pp 3639–3644
110. Peternel L, Tsagarakis N, Caldwell D, Ajoudani A (2018) Robot adaptation to human physical fatigue in human-robot co-manipulation. Auton Robot 42(5):1011–1021
111. Maclachlan RA, Becker BC, Tabarés JC, Podnar GW, Lobes LA Jr, Riviere CN (2011) Micron: an actively stabilized handheld tool for microsurgery. IEEE Trans Robot 28:195–212
112. Lawitzky M, Mörtl A, Hirche S (2010) Load sharing in human-robot cooperative manipulation. In: 19th international symposium in robot and human interactive communication. IEEE, pp 185–191

113. Donner P, Christange F, Lu J, Buss M, Lu J, Buss M (2017) Cooperative dynamic manipulation of unknown flexible objects joint energy injection based on simple pendulum fundamental dynamics. Int J Soc Robot 9(4):575–599
114. Campa R, De la Torre H (2009) Pose control of robot manipulators using different orientation representations: a comparative review. Proc Am Control Conf 1:2855–2860
115. Tsumugiwa T, Yokogawa R, Hara K (2001) Variable impedance control with regard to working process for man-machine cooperation-work system. In: 2001 IEEE/RSJ international conference on intelligent robots and systems, proceedings. vol 3. IEEE, pp 1564–1569
116. Al-Jarrah OM, Zheng YF (1996) Arm-manipulator coordination for load sharing using compliant control. In: Proceedings of IEEE international conference on robotics and automation, vol 2. IEEE, pp 1000–1005
117. Maeda Y, Hara T, Arai T (2001) Human-robot cooperative manipulation with motion estimation. In: Proceedings 2001 IEEE/RSJ international conference on intelligent robots and systems. vol 4, pp 2240–2245
118. Ivaldi S, Fritzsche L, Babič J, Stulp F, Damsgaard M, Graimann B, Bellusci G, Nori FS (2017) Anticipatory models of human movements and dynamics : the roadmap of the AnDy project. In: 5th international digital human modeling symposium
119. Dumora J, Geffard F, Bidard C, Fraisse P (2013) Towards a robotic partner for collaborative manipulation. In: HRI-workshop on collaborative manipulation, pp 1–6
120. Karayiannidis Y, Smith C, Kragic D (2014) Mapping human intentions to robot motions via physical interaction through a jointly-held object. In: The 23rd IEEE international symposium on robot and human interactive communication. IEEE, pp 391–397
121. Roveda L, Castaman N, Ghidoni S, Franceschi P, Boscolo N, Pagello E, Pedrocchi N. (2018) Human-robot cooperative interaction control for the installation of heavy and bulky components. In: 2018 IEEE international conference on systems, man, and cybernetics (SMC). IEEE, pp 339–344
122. Duchaine V, Gosselin CM (2007) General model of human-robot cooperation using a novel velocity based variable impedance control. In: EuroHaptics conference, 2007 and symposium on haptic interfaces for virtual environment and teleoperator systems. World Haptics 2007. Second Joint. IEEE, pp 446–451
123. Ikeura R, Monden H, Inooka H (1994) Cooperative motion control of a robot and a human. In: Proceedings of 1994 3rd IEEE international workshop on robot and human communication. IEEE, pp 112–117
124. Ikeura R, Inooka H (1995) Variable impedance control of a robot for cooperation with a human. Proc - IEEE Int Conf Robot Autom 3:3097–3102
125. Ikeura R, Moriguchi T, Mizutani K (2002) Optimal variable impedance control for a robot and its application to lifting an object with a human. In: Proceedings. 11th IEEE international workshop on robot and human interactive communication. IEEE, pp 500–505
126. Hayashibara Y, Takubo T, Sonoda Y, Arai H, Tanie K (1999) Assist system for carrying a long object with a human-analysis of a human cooperative behavior in the vertical direction. In: Proceedings 1999 IEEE/RSJ international conference on intelligent robots and systems. human and environment friendly robots with high intelligence and emotional quotients (Cat. No. 99CH36289), vol 2. IEEE, pp 695–700
127. Noohi E, Žefran M, Patton JL (2016) A model for human-human collaborative object manipulation and its application to human-robot interaction. IEEE Trans Rob 32(4):880–896
128. Crespo M, Mujica M, Benoussaad M, Junco S (2021) Robust control in task space for redundant manipulator. In: 2021 XIX workshop on information processing and control (RPIC), pp 1–6
129. Kronander K, Billard A (2016) Stability considerations for variable impedance control. IEEE Trans Rob 32(5):1298–1305
130. Grafakos S, Dimeas F, Aspragathos N (2017) Variable admittance control in pHRI using EMG-based arm muscles co-activation. In: 2016 IEEE international conference on systems, man, and cybernetics, SMC 2016 - conference proceedings, pp 1900–1905

131. Ferraguti F, Talignani Landi C, Sabattini L, Bonfe M, Fantuzzi C, Secchi C (2019) A variable admittance control strategy for stable physical human-robot interaction. Int J Robot Res 38(6):747–765
132. Roveda L, Haghshenas S, Prini A, Dinon T, Pedrocchi N, Braghin F, Tosatti LM (2018) Fuzzy impedance control for enhancing capabilities of humans in onerous tasks execution. In: 2018 15th international conference on ubiquitous robots (UR). IEEE, pp 406–411
133. Aydin Y, Tokatli O, Patoglu V, Basdogan C (2017) Fractional order admittance control for physical human-robot interaction. In: 2017 IEEE world haptics conference (WHC), no. June, (Fürstenfeldbruck (Munich), Germany), pp 257–262
134. Kim W, Peternel L, Lorenzini M, Babič J, Ajoudani A (2021) A human-robot collaboration framework for improving ergonomics during dexterous operation of power tools. Robot Comput-Integr Manuf 68:102084
135. Doncieux S, Chatila R, Straube S, Kirchner F (2022) Human-centered AI and robotics. AI Perspect 4(1):1
136. Khatib O, Brock O, Chang K-S, Ruspini D, Sentis L, Viji S (2004) Human-centered robotics and interactive haptic simulation. Int J Robot Res 23:167–178
137. Huo Y, Li X, Zhang X, Sun D (2021) Intention-driven variable impedance control for physical human-robot interaction. In: 2021 IEEE/ASME international conference on advanced intelligent mechatronics (AIM), pp 1220–1225
138. Ravichandar HC, Trombetta D, Dani AP (2018) Human intention-driven learning control for trajectory synchronization in human-robot collaborative tasks. IFAC-PapersOnLine 51(34):1–7. 2nd IFAC conference on cyber-physical and human systems CPHS
139. Nikolaidis S, Hsu D, Srinivasa S (2017) Human-robot mutual adaptation in collaborative tasks: models and experiments. Int J Rob Res 36:618–634
140. Vianello L, Ivaldi S, Aubry A, Peternel L (2022) The effects of role transitions and adaptation in human-robot collaboration. working paper or preprint
141. An X, Cui Y, Sun H, Shao Q, Zhao H (2023) Active-cooling-in-the-loop controller design and implementation for an SMA-driven soft robotic tentacle. IEEE Trans Rob 39(3):2325–2341
142. Ferraguti F, Secchi C, Fantuzzi C (2013) A tank-based approach to impedance control with variable stiffness. In: 2013 IEEE international conference on robotics and automation, pp 4948–4953
143. Schindlbeck C, Haddadin S (2015) Unified passivity-based cartesian force/impedance control for rigid and flexible joint robots via task-energy tanks. In: 2015 IEEE international conference on robotics and automation (ICRA). IEEE, pp 440–447
144. Mujica M, Donaire A, Benoussaad M, Fourquet JY (2021) Impedance control for a flexible robot enhanced with energy tanks in the port-hamiltonian framework. In: 2021 IEEE/RSJ international conference on intelligent robots and systems (IROS), pp 9283–9289

Mourad Benoussaad is an associate professor of the University of Technology of Tarbes since 2014. He is a member of LGP Laboratory (Laboratoire Génie de Production) in MAVRICS research group. His main research focus on Physical Human-Robot Interaction (pHRI) for the human robot collaboration (cobotics). In 2023, he received the habilitation to conduct research (HDR) from the University of Toulouse III—Paul Sabatier in this field. His research also interest on modeling, estimation and control of human musculoskeletal system in this context. Mourad Benoussaad is an author and co-author of more than 30 papers in international journals and conferences in Robotics and biomedical engineering. He received a Ph.D in Robotics and Control Theory from Science University of Montpellier (France) in 2009, and achieved several postdocs at: The University of Tokyo (2010–2011), Heidelberg University (2012–2013), INRIA Montpellier (2013–2014).

Human-Centered Edge AI and Wearable Technology for Workplace Health and Safety in Industry 5.0

Tho Nguyen, Dac Hieu Nguyen, Quoc-Thông Nguyen, Kim Duc Tran, and Kim Phuc Tran

Abstract This research explores the integration of human-centered edge artificial intelligence (AI) and wearable technology to enhance workplace health and safety within Industry 5.0. It highlights the importance of real-time monitoring and analysis facilitated by wearable devices equipped with sensors to measure physiological and environmental parameters, which help prevent hazards and accidents. Leveraging the Industrial Internet of Things (IIoT), these wearable technologies continuously track worker health and environmental conditions, promoting proactive hazard prevention and improving workplace efficiency. The role of edge AI is emphasized for its ability to enable immediate decision-making and reduce latency by processing data closer to its source, thereby enhancing worker safety and productivity while addressing ethical concerns related to privacy and security. The potential of human-centered edge AI and wearable technology to foster a collaborative and sustainable industrial environment is significant, though challenges such as limited computational resources and battery life constraints require ongoing research and development. This comprehensive analysis underscores the transformative impact of these technologies on workplace health and safety and the necessity for innovative solutions to overcome current limitations, aiming to improve worker well-being and boost industrial efficiency and productivity.

Keyword Wearable technology · Human-centered edge AI · Smart manufacturing · Industry 5.0

T. Nguyen · D. H. Nguyen · Q.-T. Nguyen · K. D. Tran (✉) · K. P. Tran
International Chair in DS & XAI, International Research Institute for Artificial Intelligence and Data Science, Dong A University, Danang, Vietnam
e-mail: ductk@donga.edu.vn

K. P. Tran
Univ. Lille, ENSAIT, ULR 2461 - GEMTEX - Génie et Matériaux Textiles, 59000 Lille, France

D. H. Nguyen
Institute of Artificial Intelligence and Data Science, Thuyloi University, Hanoi, Vietnam

K. P. Tran (ed.), *Artificial Intelligence for Safety and Reliability Engineering*, Springer Series in Reliability Engineering,
https://doi.org/10.1007/978-3-031-71495-5_8

1 Introduction

Workplace health and safety is the prevention and limitation of hazards, accidents, and other adverse effects in a workplace environment and is an essential rule in industry. Hazard is the potential of a substance, activity, or process to cause harm and accident or occupational disease [1]. The occurrence of industrial accidents, apart from causing fatalities and disabilities, is costly to the employer in terms of productivity and the family's economic burden. Tompa et al. [2] assessed economic burden of occupational diseases and work accidents in five EU countries has shown that Finland has the highest incidence (72%), then Poland (45%), and the highest economic burden of two countries based on GPD is 2.9–10.2%. There are three risk assessment methods of safety and health in the work environment: self-reports, observational, and direct measurements [3]. In Industry 4.0, direct measurements are mainly concerned with and studied with more reliable, feasible, and high-accuracy direct measurements that rely on sensors from wearable devices integrated directly into the workers. These sensors record the work activity's exposure variables.

In Industry 4.0, Artificial Intelligence (AI) applications are studied and implemented for managing the potential risks and hazards to workers. Moreover, AI also strengthens autonomous operation by increasing productivity, streamlining processes, and integrating value chains [4]. However, the unfavorable side is high automation and independence, which leads to only using workers for cognitive tasks at the workplace, such as controlling and operating machines. As the role of humans weakens in Industry 4.0 and the threats from the rapid growth of AI loom large, humans are on the path of unsustainable and unsafe development [5]. To address this, the researchers and the European Commission have proposed integrating human elements into cyber-physical systems, known as Human Cyber-Physical Systems (HCPS). This proposal is specifically designed for Industry 5.0, where humans are placed at the forefront, fostering the development of sustainable, safe, and responsible AI technology [6, 7].

The era of Industry 5.0, also known as the "Human-Centric Industrial Revolution", emphasizes the collaboration between humans and advanced technologies through cognitive abilities, problem-solving skills, and emotional intelligence from humans, while machines handle repetitive and physically demanding tasks. This collaboration aims to create a work environment that leverages the strengths of both humans and machines to drive innovation, productivity, and sustainable growth. In Industry 5.0, Edge AI provides real-time intelligence, enables immediate decision-making, and empowers edge devices to process and analyze data autonomously. Edge AI allows for faster insights, predictive maintenance, intelligent automation, and personalized user experiences. Besides, Edge Computing involves the processing and analysis of data closer to the data source, reducing latency and eliminating the need for extensive data transmission to centralized servers. The future of industry involves autonomous operation by installing new sensors and utilizing AI and machine learning to handle vast amounts of data near the source. The trend is to move the handling of the vast amount of data from sensors installed in machines and devices close to the sensors. This way, data communication can be heavily reduced, while the latency of decisions based on data can be minimal. Therefore, moving the advanced data analysis capabilities from the cloud to the edge is the prerequisite factor for the success of AI applications in Industry 5.0.

This chapter aims to investigate the potential benefits of combining human-centered artificial intelligence (HCAI) and wearable technologies to improve workplace health and safety in Industry 5.0. Emphasizing the significance of continuously monitoring and preventing hazards in real time using wearable devices equipped with sensors that measure physiological and environmental data. The research demonstrates how the Industrial Internet of Things (IIoT) may be utilized to prevent dangers and enhance productivity through continuous tracking. Overall, it provides a comprehensive analysis of the benefits, challenges, and future perspectives of integrating advanced technologies to enhance workplace health and safety.

The rest of this chapter is structured as follows: We begin with "Wearable Industrial Internet of Things" in Sect. 2. This section will discuss how wearable technology can be integrated within the Industrial Internet of Things (IIoT) framework to enhance workplace safety and productivity through real-time monitoring and data analysis. Next, in Sect. 3, we will explore "Human-Centered AI for Workplace Health and Safety," examining how incorporating human-centered artificial intelligence (AI) can further improve workplace safety protocols and health outcomes. This section will cover the principles of human-centered AI, focusing on its ability to provide personalized, real-time insights and decision-making capabilities, thereby reducing latency and enhancing worker safety. Following that, in Sect. 4, titled "Difficulties, Challenges, and Perspectives for Application of Human-Centered Edge AI and Wearable Technology Techniques for Smart Manufacturing in Industry 5.0," we will address the practical challenges and limitations associated with implementing these advanced technologies. Lastly, the chapter will conclude with "Concluding Remarks" in Sect. 5 summarizing the key insights discussed throughout the chapter. This section will emphasize the transformative potential of integrating human-centered edge AI and wearable technology in enhancing workplace health and safety while acknowledging the existing challenges and the need for innovative solutions to realize this potential fully.

2 Wearable Industrial Internet of Things

Wearable technology encompasses small electronic devices with wireless communications capabilities integrated into gadgets, accessories, or clothing. These devices are attached to either the human body or clothing to gather and track biological and activity data from users [8, 9]. Wearable devices are utilized to track human activity recognition for risk assessment purposes, and wearable sensor networks are seen as an effective way to improve work efficiency by assisting human activities and enhancing workers' well-being [10, 11]. Moreover, wearable devices allow constant tracking of individual workers and the environment [12]. Connected worker solutions provide contextual information and decision support, along with wearable devices specifically designed for ensuring workplace safety during activities such as lifting, maintaining ergonomic best practices, identifying hazards, monitoring sleep patterns, managing fatigue, and addressing issues related to heat and cold exposure [12, 13]. These wearable devices also enhance workplace productivity through applications such as asset tracking, augmented reality, gesture and motion control, brain wave sensing, and managing work-related stress [12].

Traditional wearable devices face portability issues and limited memory, processing, and network connectivity due to multitasking constraints. The IoT is utilized to address these challenges [14]. However, the real-world impact of this technology becomes even more apparent when we integrate wearable technology with the Industrial Internet of Things (IIoT). This integration has shown significant potential in various industries, and this section introduces the practical applications of the Wearable Industrial Internet of Things for the safety and well-being of workers in the workplace.

In manufacturing, the Wearable Internet of Things (WIoT) represents a distinct category of enabling technology that leverages data to enhance efficiency and productivity. WIoT devices hold substantial promise in quality assurance, traceability, and the optimization of supply chain operations. Sensors integrated into the apparatus and attire of manufacturing personnel monitor their output, enabling the recalibration and refinement of performance as dictated by operational demands. These IoT functionalities integrate employees and their associated machinery into a cohesive system. Firms are devising manageable systems capable of overseeing all production activities in real time to augment process efficiency. In the fiercely competitive manufacturing sector, prioritizing efficiency to facilitate continuous improvement and expansion is crucial. Several enterprises have realized the benefits of wearable technologies that propel them towards realizing a 'smart factory' characterized by optimized time utilization and minimized operational downtimes. For example, employees use smart IoT glasses in the industrial food sector to manage overdue food storage and present notifications to relevant personnel. Data findings are delivered to cloud storage, where management may examine them instantly and take the appropriate corrective action to solve issues. This benefit reduces the time it takes to handle important food safety issues and eliminates the need to dig through mountains of paperwork to find problems. Additionally, employees can spend less time collecting data via IoT wearables (automating the recording process) than paper systems (writing data on paper). The most important feature is that wearable IoT sensors work well in anomaly detection of food if their parameters fall out of the safe set range. Some types of IoT wearables applied in industrial food safety are Smart Eyewear (visualization, communication), Wearable Cameras (photographic or audio-video evidence), and Smart Swabs (real-time bacterial activity) [15].

Biomechanical exposures during physical activity by intensity (e.g., load magnitudes, extent of non-neutral postures), repetition (frequency or number of force exertions and motions), and duration (the time the physical activity is performed, sustained non-neutral postures) [13, 16]. To achieve accurate assessment results of work-related and occupational diseases, it is necessary to monitor and observe for a long time, meaning that data needs to be collected continuously over a long period. A comprehensive description of the parameters that need to be evaluated and the wearable technology method allows this problem to be solved. Accurate, reliable, and cost-effective quantification of real-time biomechanical exposures in occupational settings remains an enduring pursuit in Industry 5.0.

In the construction field, performing construction tasks often require workers to assume awkward postures. The repetition of awkward postures coupled with long periods of exposure will result in musculoskeletal disorders (MSDs) such as chronic back pains and over-exertion. In the United States, construction-related MSDs account for 30% of workplace injuries [17], and construction workers have an 18% higher rate of MSDs compared to all private industries [18]. The Wearable Internet of Things is considered a potential solution for workers' safety and productivity in the construction field.

A notable aspect of the manufacturing area is a protocol for third-party confirmation before any worker accesses devices. Therefore, wearable technology can remotely restrict, allow, and location-track worker activities. In 2018, [19] suggested a prototype incorporating textile electrodes, motion sensors, and real-time data processing through a mobile application for autonomous risk assessment of physical activity. [20] used sensors placed on clothes to alert nurses when they are in the wrong posture at work.

In the industrial sector and firefighting field, wearable sensors are also applied in the undergarment for monitoring physiological extremes [21]. The appearance of voice-assisted tech (Google's Assistant, Apple's Siri, and Samsung's Bixby) shows the huge potential when integrating voice assistants into wearable IoT systems. This technology is more advantageous than the smart devices mentioned. Imaginatively, a worker can control a machine using a wearable IoT device with a voice assistant instead of working with the machine by hand. In addition, the time for movement in a company is minimized when staff can remotely order machines, which helps staff work with greater productivity. With the development of voice assistants, wearable IoT devices are poised for rapid development, driving us closer to Industry 5.0. The higher level of voice assistants and the virtual entity is the Human Digital Twin (HDT), the digital representation of the real-world human or real-world twin [22]. In the Industry 4.0 context, a digital twin that comprises wearable devices, IoT networks, and AI is developed to boost workforce productivity [23]. In addition, it can simulate the future behavior of operators, reduce physical workload, and increase the well-being of operators. Towards Industry 5.0, Human-Cyber-Physical Systems (HCPS) are most concerned with optimized manufacturing levels. Wearable technologies connect humans and machines and supply rich information about the operators, surrounding environment, and machines [24, 25]. Besides concerns related to efficient productivity, many companies use IoT-enabled wearables to improve safety and prevent worker hazards. Health monitoring and management for manufacturing workers in adverse conditions is for their safety and cost efficiency [26]. Wearable technology is applied for ergonomic risk assessment in manufacturing, such as identifying work-related musculoskeletal disorders via measuring and assessing the exposure to awkward postures and sEMG sensors, which provide the measurement of muscle activity parameters related to the force applied while performing work activities.

Until now, studies and practical applications of wearable edge AI devices remain fragmented. Yang et al. [27] has experimented with a wearable edge AI platform to collect physiological signals for emotion recognition classification (happiness, anger, sadness) via Electroencephalogram (EEG), electrocardiogram (ECG), and photoplethysmogram (PPG) data with a convolution neural network (CNN), resulting in an average accuracy of 76.8%. Though the obtained result is not high, the experiment has proved the high potential of edge AI wearable gadgets in the future. The findings also contribute to healthcare monitoring devices to early detect workers' occupational diseases that may link to cardiovascular disorders [27].

3 Human-Centered AI for Workplace Health and Safety

Human-Centered Artificial Intelligence (HCAI) is a system that constantly learns and improves models via recognition of experience and behaviors of human input, which aims to understand human language [28]. An HCAI approach to designing and developing systems supports human self-efficacy, encourages creativity, clarifies responsibility, and facilitates social participation. These foundational principles can help guide designers toward vital technical goals, including privacy, security, fairness, reliability, safety, and trustworthiness. HCAI ensures that AI algorithms are technically proficient and align with principles such as fairness, accountability, and societal benefits [29].

Artificial Intelligence has proven its success in prediction and classification tasks; however, the challenges in HCAI are analyzing the massive volume of data in the cloud, computing velocity, transmission latency, and privacy. Therefore, a potentially more feasible solution is to move the AI computing at the core of the network cloud to the network edges [30, 31]. This solution can be called Edge AI or Edge Intelligence (EI). The core idea of the EI model is to offload the data processing, storage, and computing operations originally required by the cloud to the edge of the network near terminal devices. Edge AI does not have centralized computing resources, which reduces data transmission time and device response times, reduces the pressure on network bandwidth, reduces the data transmission cost, and achieves decentralization [32]. Edge Intelligence refers to doing AI computations near the users at the network's edge instead of a centralized location like a cloud service provider's data center [33]. Until 2024, most AI processes are currently performed in cloud-based centers due to requiring substantial computing capacity. However, massive and complex data leads to downsides, such as downtime or a significant service slowdown. Edge AI devices eliminate these problems by not needing to be connected to the Internet to function; they can process data, train models, and make decisions independently, an integral part of edge computing devices. This enables users to save time by aggregating data and serving users without communicating with other physical locations. An edge device practically means any hardware not located in some central cloud (i.e., not in a server rack somewhere in a large data center) with sufficient computing power to do necessary calculations. Therefore, training AI models at the edge will remarkably enhance the range and computational speed of IoT-based devices in industries [34]. Edge computing is suitable for applications in technologies that use a vast amount of data, such as autonomous vehicles, smart factories, and mobile applications, which are currently being developed and are important for real-time processing. One example would be the Internet of Vehicles (IoV) [35], which has requirements of high speed and low latency for data processing, analyzing, and returning results [36].

Integrating wearable devices in workplace health and safety, particularly in the context of the construction industry, presents significant opportunities for enhancing safety and health management practices. Wearable Internet of Things (WIoT) technologies provide real-time monitoring capabilities such as body temperature, heart rate, stress levels, and breathing rates [37]. Moreover, adopting wearable sensing devices (WSDs) is increasingly enhancing worker safety and well-being in diverse industries [38]. By leveraging these devices, organizations can track workers' well-being, ensure safe working environments, and mitigate health risks associated with physically demanding and hazardous work settings. Wearable devices equipped with biosensors and AI systems enable data collection and analysis, facilitating the monitoring of employees' health and safety in various industries [39]. In addition, the COVID-19 pandemic has emphasized the importance of wearable devices integrated with AI for symptom tracking and outcome prediction, emphasizing the evolving role of these technologies in healthcare [40]

Adopting AI technologies provides organizations with predictive insights, real-time monitoring capabilities, and risk mitigation strategies that reduce accidents and hazards and enable a more proactive approach to safeguarding the workforce [41]. AI can serve as a valuable tool for addressing the hazards faced by workers and promoting health in the workplace [42]. Furthermore, the development of AI in workplaces is transforming work practices and skill requirements, highlighting the need for a deeper understanding of the implications of AI on labor markets and the future of work [43]. Researchers have demonstrated the practical application and effectiveness of AI-driven strategies in addressing workplace issues such as bullying, leading to significant reductions in incidents, improved employee well-being, and enhanced cultural awareness [44]. Besides developing machine learning algorithms, sophisticated sensors, and robotics are progressively employed to oversee and enhance occupational health and safety (OHS) [45]. These technologies can identify dangerous situations, anticipate possible dangers, and issue immediate notifications to prevent accidents and injuries. By leveraging AI technologies effectively, organizations can enhance worker safety, mitigate risks, and promote a culture of safety in the workplace. However, organizations need to address ethical considerations, ensure transparency, and adopt human-centric approaches to AI to maximize its benefits while safeguarding the well-being of employees.

4 Difficulties, Challenges, and Perspectives for Application of Human-Centered Edge AI and Wearable Technology Techniques for Smart Manufacturing in Industry 5.0

Integrating Human-Centered Edge AI and wearable technology into smart manufacturing offers significant benefits such as improved worker safety and productivity. However, this integration comes with challenges and this section explores these difficulties and future perspectives for effectively utilizing these advanced technologies in Industry 5.0.

Human-Centered Edge AI is a concept in artificial intelligence (AI) and edge computing focusing on developing and deploying AI systems that operate close to the end user. This ensures a better user experience by enhancing interaction and meeting specific needs. These systems prioritize security and privacy by processing data locally, minimizing the risks of transmitting sensitive data over networks. Additionally, Human-Centered Edge AI delivers high real-time performance, enabling AI applications to respond quickly and accurately since data is processed at the collection point [46, 47]. This is particularly useful in fields like healthcare, where smart wearable devices can collect patient health data and analyze it immediately to support diagnosis and disease management. Integrating edge AI also helps personalize healthcare services and protects patient privacy [48]. Thus, Human-Centered Edge AI enhances the efficiency and availability of AI systems and ensures they are user-friendly and secure for the end-users.

4.1 Ethical Considerations and User Experiences

The ethical concerns of HCAI are substantial, due to an almost unverified training dataset, which may unintentionally reinforce biases in the data and provide discriminating results. Developers attempt to focus on avoiding harm by ensuring ethics, privacy, and security are integral to AI design, such as ethical AI frameworks proposed by Floridi et al. [49] and the European Commission [50], emphasize these principles but are often seen as too abstract and difficult to apply in practice. However, according to the survey of Bingley et al. [51], positive user experiences were characterized by the social impact of AI more than knowing AI, users were concerned about being understood by AI. Highlighting the gap between developer intentions and user expectations. In the future, researchers need more practical and user-centric approaches to ethical AI implementation, ensuring that AI systems are not only fair and unbiased but also align closely with user expectations and societal values.

4.2 Privacy and Security Concerns

Wearable technologies collect vast amounts of personal data, raising significant privacy and security concerns. Using outdated bluetooth protocols or sending of information via unencrypted channels (e.g., HTTP instead of HTTPS) in wearable devices further exacerbates security risks. Moreover, many wearable devices use static MAC addresses instead of dynamic MAC addresses, enabling attackers to track users' movements and activities [52]. To address these challenges, Echenim et al. [53] developed a novel framework for evaluating privacy policies governing the data and information utilized by wearable devices to ensure that these IoT devices guarantee compliance with NISTIR 8228 regulations [54]. The methodology utilizes Natural Language Processing, Semantic Web concepts, and the transformation of textual documents into graph-based ontologies to effectively extract expectations, privacy risk mitigation areas, and associated regulations. However, security and privacy require comprehensive solutions to ensure the protection of user data. Researchers are continually developing advanced privacy-preserving methods, robust compliance frameworks, and innovative security measures to address these ongoing challenges and enhance the overall safety of wearable IoT devices.

4.3 Wearable Technology Limitations

The hardware of end devices is usually limited and fixed, whereas the hardware of cloud servers can be easily upgraded to have more memory, powerful CPU cores, and GPU cards that enable push machine learning inference to ultralow-power devices. Hardware and software design are required to unleash the full potential of on-device inference. Advances in computing infrastructure have significantly contributed to this capability. Powerful distributed computational power is required to run AI at the edge, and recent advances in highly parallel GPUs have been adapted to execute neural networks efficiently. These advancements facilitate the deployment of complex AI models directly on edge devices, enabling real-time processing and decision-making while minimizing latency and bandwidth usage.

Wearable devices frequently face constraints related to battery longevity, which can hinder their continuous operation in the industrial environment. Not only focusing on optimizing the complexity but also developing a power-efficient algorithm, low-power design, and energy-harvesting techniques, that can significantly prolong the battery life of wearables. Many researchers have attempted to explore the energy-efficient design, such as a wearable pulse oximeter assembled in a 3D ring-like geometry presented by Magno et al. [55], it achieves self-sustainability by exploiting efficient power management, solar energy harvesting, and ultra-low power processing in a multi-core microcontroller, combining onboard processing to monitor blood oxygenation and the transmission of only relevant information via a Bluetooth low-energy (BLE) interface. Recently, Ali et al. [56] presents a detailed analysis of current advancements in energy harvesting systems for wearable technologies. By leveraging these advancements, wearable technology can achieve longer operational duration, enhancing user experience and usability.

4.4 Deploying AI in Wearable Devices

Enhancing hardware is not sufficient, it is also essential to optimize AI models to enable the deployment of AI on wearable devices. Wearable devices embedded with AI have the potential to revolutionize various sectors, particularly healthcare, by enabling personalized health monitoring systems, chronic disease management, and remote patient monitoring [57–59]. Challenges for AI models on wearable devices include limited computational power, memory, and battery life, along with the need for real-time processing and data privacy. To address these, model compression techniques such as pruning and quantization reduce the size and computational demands of deep learning models [60]. Nonetheless, the limited resources of these devices are a hurdle, and ongoing research is expected to continue in this field.

5 Concluding Remarks

Integrating Human-Centered Edge AI and wearable technology into workplace health and safety systems marks a significant advancement, especially within Industry 5.0. This paradigm shift from traditional methods to technology-driven solutions emphasizes real-time data processing, immediate decision-making, and enhanced worker safety and productivity. Wearable technology continuously monitors workers' health and environmental conditions, offering a proactive approach to hazard prevention. Within Industrial Internet of Things (IIoT) frameworks, these devices detect risks and improve workplace efficiency by processing data close to its source, reducing latency and improving response times. Human-Centered Edge AI focuses on user-centric design, privacy, and real-time performance, particularly benefiting high-stakes environments like healthcare and manufacturing. By processing data locally, these systems mitigate privacy concerns and enhance security. This approach fosters a collaborative environment in Industry 5.0, where human cognitive skills and machine efficiency drive innovation, productivity, and sustainable growth. However, challenges such as limited computational resources of edge devices, battery life constraints, and security vulnerabilities require ongoing research and development. Addressing these involves creating lightweight, power-efficient AI models and robust security frameworks. In conclusion, integrating Human-Centered Edge AI and wearable technology in workplace safety enhances worker well-being and boosts industrial efficiency and productivity. Innovative solutions must focus on human-centric design, ethical concerns, privacy, and real-time performance as technology evolves.

References

1. Rout B, Sikdar B (2017) Hazard identification, risk assessment, and control measures as an effective tool of occupational health assessment of hazardous process in an iron ore pelletizing industry. Indian J Occup Environ Med 21:56–76
2. Tompa E, Mofidi A, van den Heuvel S, van Bree T, Michaelsen F, Jung Y, Porsch L, van Emmerik M (2021) Economic burden of work injuries and diseases: a framework and application in five European union countries. BMC Public Health 21:1–10
3. David GC (2005) Ergonomic methods for assessing exposure to risk factors for work-related musculoskeletal disorders. Occup Med 55:190–199
4. Cebulla A, Szpak Z, Howell C, Knight G, Hussain S (2023) Applying ethics to AI in the workplace: the design of a scorecard for Australian workplace health and safety. AI & Soc 38:919–935
5. Luthra S, Mangla SK (2018) Evaluating challenges to industry 4.0 initiatives for supply chain sustainability in emerging economies. Process Saf Environ Protect 117:168–179
6. Commission E (2021) Industry 5.0: Human-centric, sustainable and resilient
7. Wang D, Zhang Y (2023) Social edge computing: empowering human-centric edge computing. Springer, Learning and Intelligence
8. Burmaoglu S, Trajkovik V, Tutukalo TL, Yalcin H, Caulfield B (2018) Evolution map of wearable technology patents for healthcare field. Wearable Technology in Medicine and Health Care, pp 275–290
9. Luczak T, Burch R, Lewis E, Chander H, Ball J (2020) State-of-the-art review of athletic wearable technology: what 113 strength and conditioning coaches and athletic trainers from the usa said about technology in sports. Int J Sports Sci Coach 15:26–40
10. Khakurel J, Pöysä S, Porras J (2016) The use of wearable devices in the workplace-a systematic literature review. In: Smart objects and technologies for social good: second international conference, GOODTECHS 2016, Venice, Italy, Proceedings 2. Springer, pp 284–294. Accessed from November 30–December 1, 2016
11. Stefana E, Marciano F, Rossi D, Cocca P, Tomasoni G (2021) Wearable devices for ergonomics: a systematic literature review. Sensors 21:777
12. Patel V, Chesmore A, Legner CM, Pandey S (2022) Trends in workplace wearable technologies and connected-worker solutions for next-generation occupational safety, health, and productivity. Adv Intell Syst 4:2100099
13. Sabino I, do Carmo Fernandes M, Cepeda C, Quaresma C, Gamboa H, Nunes IL, Gabriel AT (2024) Application of wearable technology for the ergonomic risk assessment of healthcare professionals: a systematic literature review. Int J Ind Ergonom 100:103570
14. Nguyen T, Tran KD, Raza A, Nguyen QT, Bui HM, Tran KP (2023) Wearable technology for smart manufacturing in industry 5.0. Springer International Publishing, Cham, pp 225–254
15. Bader F, Jagtap S (2020) Internet of things-linked wearable devices for managing food safety in the healthcare sector. In: Wearable and implantable medical devices. Elsevier, pp 229–253
16. Andersen JH, Evanoff BA, Descatha A (2023) Biomechanical hazards at work and adverse health using job-exposure matrices. In: Handbook of life course occupational health. Springer, pp 1–14
17. Zhao J, Obonyo E, Bilén SG (2021) Wearable inertial measurement unit sensing system for musculoskeletal disorders prevention in construction. Sensors 21:1324
18. Okpala I, Nnaji C, Ogunseiju O, Akanmu A (2022) Assessing the role of wearable robotics in the construction industry: potential safety benefits, opportunities, and implementation barriers. Automation and robotics in the architecture, engineering, and construction industry, pp 165–180
19. Yang L, Lu K, Diaz-Olivares JA, Seoane F, Lindecrantz K, Forsman M, Abtahi F, Eklund JA (2018) Towards smart work clothing for automatic risk assessment of physical workload. IEEE Access 6:40059–40072
20. Bootsman R, Markopoulos P, Qi Q, Wang Q, Timmermans AA (2019) Wearable technology for posture monitoring at the workplace. Int J Hum Comput Stud 132:99–111

21. Salim F, Prohasky D, Belbasis A, Houshyar S, Fuss FK (2014) Design and evaluation of smart wearable undergarment for monitoring physiological extremes in firefighting. In: Proceedings of the 2014 ACM international symposium on wearable computers: adjunct program, pp 249–254
22. Miller ME, Spatz E (2022) A unified view of a human digital twin. Human-intelligent systems integration, pp 1–11
23. Montini E, Bettoni A, Ciavotta M, Carpanzano E, Pedrazzoli P (2021) A meta-model for modular composition of tailored human digital twins in production. Proc CIRP 104:689–695
24. Zhou J, Zhou Y, Wang B, Zang J (2019) Human-cyber-physical systems (HCPSS) in the context of new-generation intelligent manufacturing. Engineering 5:624–636
25. Sun S, Zheng X, Gong B, Garcia Paredes J, Ordieres-Meré J (2011) Healthy operator 4.0: a human cyber–physical system architecture for smart workplaces. Sensors 20(2020)
26. Xu X, Zhong M, Wan J, Yi M, Gao T (2016) Health monitoring and management for manufacturing workers in adverse working conditions. J Med Syst 40:1–7
27. Yang CJ, Fahier N, He CY, Li WC, Fang WC (2020) An AI-edge platform with multimodal wearable physiological signals monitoring sensors for affective computing applications. In: 2020 IEEE international symposium on circuits and systems (ISCAS). IEEE, pp 1–5
28. Shneiderman B (2020) Human-centered artificial intelligence: three fresh ideas. AIS Trans Hum-Comput Interact 12:109–124
29. Singh BP, Joshi A (2024) Ethical considerations in AI development, IGI Global, pp 156–179
30. Zhou Z, Chen X, Li E, Zeng L, Luo K, Zhang J (2019) Edge intelligence: paving the last mile of artificial intelligence with edge computing. Proc IEEE 107:1738–1762
31. Deng S, Zhao H, Fang W, Yin J, Dustdar S, Zomaya AY (2020) Edge intelligence: the confluence of edge computing and artificial intelligence. IEEE Internet Things J 7:7457–7469
32. Ghosh AM, Grolinger K (2020) Edge-cloud computing for internet of things data analytics: embedding intelligence in the edge with deep learning. IEEE Trans Ind Inf 17:2191–2200
33. Singh R, Gill SS (2023) Edge AI: a survey. Internet of Things and Cyber-Physical Systems (2023)
34. Sodhro AH, Pirbhulal S, De Albuquerque VHC (2019) Artificial intelligence-driven mechanism for edge computing-based industrial applications. IEEE Trans Ind Inf 15:4235–4243
35. Paul A, Chilamkurti N, Daniel A, Rho S (2017) Chapter 3 - vehicular network (VN) model. In: Paul A, Chilamkurti N, Daniel A, Rho S (eds) Intelligent vehicular networks and communications. Elsevier, pp 43–75
36. Qin M, Chen L, Zhao N, Chen Y, Yu FR, Wei G (2018) Power-constrained edge computing with maximum processing capacity for IoT networks. IEEE Internet Things J 6:4330–4343
37. Okonkwo C, Awolusi I, Nnaji C (2022) Privacy and security in the use of wearable internet of things for construction safety and health monitoring. IOP Conf Ser: Earth Environ Sci 1101:092004
38. Nnaji C, Awolusi I, Park J, Albert A (2021) Wearable sensing devices: Towards the development of a personalized system for construction safety and health risk mitigation. Sensors 21:682
39. Weerts S, Naous D, Bouchikhi ME, Clavien C (2022) AI systems for occupational safety and health: from ethical concerns to limited legal solutions. Springer International Publishing, pp 499–514
40. Secara I-A, Hordiiuk D (2024) Personalized health monitoring systems: integrating wearable and AI. J Intell Learn Syst Appl 16:44–52
41. Shah IA, Mishra S (2024) Artificial intelligence in advancing occupational health and safety: an encapsulation of developments. J Occup Health 66
42. Jetha A, Bakhtari H, Rosella LC, Gignac MAM, Biswas A, Shahidi FV, Smith BT, Smith MJ, Mustard C, Khan N, Arrandale VH, Loewen PJ, Zuberi D, Dennerlein JT, Bonaccio S, Wu N, Irvin E, Smith PM (2023) Artificial intelligence and the work-health interface: a research agenda for a technologically transforming world of work. Am J Ind Med 66:815–830
43. Margaryan A (2023) Artificial intelligence and skills in the workplace: an integrative research agenda. Big Data Soc 10

44. Du J (2024) Leveraging artificial intelligence to address cross-cultural workplace bullying among international students in Australia. Int J Global Econ Manag 3:332–341
45. Shah IA, Mishra S (2024) Artificial intelligence in advancing occupational health and safety: an encapsulation of developments. J Occup Health 66:uiad017
46. Hancock B, Weddle B, Valentine M, Rahilly L (2023) Human-centered AI: the power of putting people first. McKinsey & Company
47. Chen C (2023) Ai will transform teaching and learning. let's get it right, Stanford University AI+ Education Summit. Accessed from 7 June 2023
48. Shumba A-T, Montanaro T, Sergi I, Bramanti A, Ciccarelli M, Rispoli A, Carrizzo A, De Vittorio M, Patrono L (2023) Wearable technologies and ai at the far edge for chronic heart failure prevention and management: A systematic review and prospects. Sensors 23:6896
49. Floridi L, Cowls J, Beltrametti M, Chatila R, Chazerand P, Dignum V, Luetge C, Madelin R, Pagallo U, Rossi F, Schafer B, Valcke P, Vayena E (2018) Ai4people-an ethical framework for a good AI society: opportunities, risks, principles, and recommendations. Mind Mach 28:689–707
50. Requirements of Trustworthy AI - FUTURIUM - European Commission — ec.europa.eu, https://ec.europa.eu/futurium/en/ai-alliance-consultation/guidelines/1.html, [Accessed 22-06-2024]
51. Bingley WJ, Curtis C, Lockey S, Bialkowski A, Gillespie N, Haslam SA, Ko RK, Steffens N, Wiles J, Worthy P (2023) Where is the human in human-centered AI? insights from developer priorities and user experiences. Comput Hum Behav 141:107617
52. Fúster J, Solera-Cotanilla S, Pérez J, Vega-Barbas M, Palacios R, Álvarez Campana M, Lopez G (2023) Analysis of security and privacy issues in wearables for minors. Wireless Networks
53. Echenim KU, Elluri L, Joshi KP (2023) Ensuring privacy policy compliance of wearables with IoT regulations. In: 2023 5th IEEE international conference on trust, privacy and security in intelligent systems and applications (TPS-ISA), pp 247–256
54. Boeckl K, Boeckl K, Fagan M, Fisher W, Lefkovitz N, Megas KN, Nadeau E, O'Rourke DG, Piccarreta B, Scarfone K (2019) Considerations for managing Internet of Things (IoT) cybersecurity and privacy risks. National Institute of Standards and Technology, US Department of Commerce
55. Magno M, Salvatore GA, Jokic P, Benini L (2019) Self-sustainable smart ring for long-term monitoring of blood oxygenation. IEEE Access 7:115400–115408
56. Ali A, Shaukat H, Bibi S, Altabey WA, Noori M, Kouritem SA (2023) Recent progress in energy harvesting systems for wearable technology. Energy Strategy Rev 49:101124
57. Shumba A-T, Montanaro T, Sergi I, Bramanti A, Ciccarelli M, Rispoli A, Carrizzo A, De Vittorio M, Patrono L (2023) Wearable technologies and AI at the far edge for chronic heart failure prevention and management: a systematic review and prospects. Sensors 23:6896
58. Gautam N, Ghanta S, Mueller J, Mansour M, Chen Z, Puente C, Ha Y, Tarun T, Dhar G, Sivakumar K, Zhang Y, Halimeh A, Nakarmi U, Al-Kindi S, DeMazumder D, Al'Aref S (2022) Artificial intelligence, wearables and remote monitoring for heart failure: current and future applications. Diagnostics 12:2964
59. Junaid SB, Imam AA, Balogun AO, De Silva LC, Surakat YA, Kumar G, Abdulkarim M, Shuaibu AN, Garba A, Sahalu Y, Mohammed A, Mohammed TY, Abdulkadir BA, Abba AA, Kakumi NAI, Mahamad S (2022) Recent advancements in emerging technologies for healthcare management systems: a survey. Healthcare 10:1940
60. Durmaz Incel O, Bursa S (2023) On-device deep learning for mobile and wearable sensing applications: a review. IEEE Sensors J 23:5501–5512

Safety and Reliability of Artificial Intelligence Systems

Thi Hien Nguyen, Aamir Saghir, Kim Duc Tran, Dac Hieu Nguyen, Nguyen Anh Luong, and Kim Phuc Tran

Abstract The pervasive integration of Artificial Intelligence (AI) in various facets of human life, driven by increasingly sophisticated algorithms, underscores the importance of its safety and reliability. AI's role in Industry 4.0, connecting machines and processes to solve complex issues, is paving the way for the 5.0 Industrial Revolution (5IR). This revolution addresses global challenges such as climate change, pandemics, and conflicts. Ensuring the safety and reliability of AI systems is crucial, as these technologies significantly impact society. This chapter provides an overview of AI safety and reliability, discussing major advancements, methodologies for reliability assessment, and practical examples of AI applications. It highlights the Human-Centered AI (HCAI) concept, which emphasizes aligning AI development with human values. The chapter also explores machine learning's role in enhancing AI reliability and addresses the challenges and ethical concerns associated with AI deployment. Underscore the need for ongoing research and interdisciplinary collaboration to ensure AI systems are safe, reliable, and beneficial to humanity in the evolving landscape of Industry 5.0 with applications in many fields such as healthcare, manufacturing, human resources management, etc.

T. H. Nguyen
Laboratoire AGM, UMR CNRS 8088, CY Cergy Paris Université, 95000 Cergy, France

A. Saghir
Budapest University of Technology and Economics, Institute of Mathematics, Department of Stochastics, Budapest Müegyetem rkp.3., 1111, Hungary

K. D. Tran (✉) · D. H. Nguyen · N. A. Luong · K. P. Tran
International Chair in DS & XAI, International Research Institute for Artificial Intelligence and Data Science, Dong A University, Danang, Vietnam
e-mail: ductk@donga.edu.vn

D. H. Nguyen
Institute of Artificial Intelligence and Data Science, Thuyloi University, Hanoi, Vietnam

N. A. Luong
Human Resources Administration Office, Dong A University, Danang, Vietnam

K. P. Tran
Univ. Lille, ENSAIT, ULR 2461 - GEMTEX - Génie et Matériaux Textiles, 59000 Lille, France

K. P. Tran (ed.), *Artificial Intelligence for Safety and Reliability Engineering*, Springer Series in Reliability Engineering,
https://doi.org/10.1007/978-3-031-71495-5_9

1 Introduction

We now live in a world where AI is omnipresent. We frequently question why advertisements appear on screens at the exact moment we need them to. For example, Amazon suggests buying the item; Netflix, Facebook, or TikTok seem to know which videos, movies, or series to recommend to watch. All these tools are powered by AI. As a result, the safety and reliability of AI systems are paramount. The purpose of this chapter is to go through the major contents related to the safety and reliability of AI in scientific research as well as analyze some practical examples in the world on this issue, from which we can continue to develop AI systems to serve humans better.

Many algorithms have been developed to evaluate AI reliability, for example, artificial neural networks ([71]), reliability block diagrams ([12]), fault trees, event trees ([60]), binary decision diagrams ([78]), Bayesian networks, etc. Furthermore, scientists are able to combine multiple algorithms in parallel or in batches to arrive at the most accurate and reliable solution. For example, the use of binary decision diagrams in fault tree analysis provides both an accurate and efficient means of analyzing a system ([64]). Or very recently, [13] proposes a combination Bayesian neural networks and differential evolution algorithms to evaluate the life of a product. Their results demonstrate how well this method can raise the reliability assessment's accuracy.

In the context of the Fifth Industrial Revolution (5IR), where human-machine collaboration is critical to creating customer value, a novel concept known as HCAI (Human-Centered Artificial Intelligence) has surfaced. Researchers from a variety of fields have attempted to express their opinions about HCAI by offering different definitions([66]). Therefore, synthesizing and reviewing these definitions in many different domains is necessary. A better understanding of HCAI helps scientists/engineers learn more quickly and apply HCAI to human life ([67, 74]). Section 2 will go further into this question.

To develop AI systems, of course we have to talk about Machine Learning (ML). ML involves the decision or prediction based on data. The system of ML can be understand like a very sophisticated form of statistical analysis based on data that we have. The more we feed into the system, the more it's able to give us a accurate predictions or decision. The prospect of remarkable advancements in machine learning and other algorithms piques interest and prompts significant investments in medical, manufacturing, and military advancements ([74]).

The field of ML is extremely large and has many sub-branches. The scope of a chapter cannot cover all the problems and go deep into each branch. Therefore, here we will only give an overview of the general principles and most basic implementations of ML in the Safety and Reliability AI system.

This is how the rest of the chapter is structured. Human-Centered Artificial Intelligence (HCAI) systems are introduced in brief in Sect. 2, including their definition, advantages, and areas for improvement. Section 3 makes a review of the literature pertinent to Reliability Assessment and Safety Arguments for AI Systems. The current research on the topic of machine learning in artificial intelligence is presented in Sect. 4. Section 5 discusses the obstacles and difficulties in these areas. Finally, Sect. 6 contains an overview of the study's concluding remarks.

2 Human-Centered Artificial Intelligence Systems

Artificial Intelligence (AI) is a discipline that solves problems by combine data science and computer science. In the last couple of decades, the development of AI has been rapid and strong like a storm, it has become better and dominated real-world applications. It can increase productivity, reduce cost, and quicken the pace of research and development. ([84]). However, as AI continues to evolve, it's important to remember that we need to create truly intelligent systems and human supervision of these systems. This is where Human-Centered Artificial Intelligence systems (HCAI) come in. As its name suggests, HCAI's concerns about human beings are at the top of everything. Creating a smarter AI system is not only about optimizing the model, and improving the algorithms but also about optimizing the needs, wants, and desires of humans. It is gradually becoming familiar in modernized global life.

Figure 1 gives a dialectical relationship between HCAI and humans. HCAI helps people in all aspects of their lives: save time, improves quality, reduces costs, provide health care, reduces stress, etcetera; in short, it helps people have a more convenient and simpler life. It must be noted that it also helps people understand AI systems. From this, humans will improve AI systems and their logarithms to turn back to serve humans.

Artificial intelligence (AI) models and algorithms are very good at the data they are trained on. The benefit of high-accuracy prediction is that AI systems greatly assist people in cancer diagnosis and other forecasts, such as weather or traffic jams.

However, for these smart and high-tech devices, we still cannot completely trust their predictions. Take a specific example of the case of "Watson for Oncology" (WFO) (see [47])—one of IBM's technology supercomputer programs. Although WFO cannot currently replace oncologists, research has shown that it can help doctors work more efficiently in the clinic by supporting them. However, in order to become more helpful, WFO must be trained to recognize the unique characteristics of each patient's region. Therefore, many researchers have tried to create algorithm and applications to improve the reliability of AI. For example, [72] propose fifteen suggestions to improve the dependability, security, and credibility of HCAI systems. As an additional illustration, there are numerous apps available today that enable

Human-Centered Artificial Intelligence $\rightleftarrows$ Human

Fig. 1 Relation between HCAI and Human

people to converse in any language without any prior language knowledge. However, acquiring a language also entails learning about the culture of the speakers of that language. Cultural awareness aids in avoiding misunderstandings; artificial intelligence cannot do this. While AI can help us learn languages more quickly, it cannot take our place.

Every day we are faced with a lot of information coming from fake deep. The catastrophe risks potential when an AI system learns from face and voice data systems. With a few simple steps, hackers can use face/voice recognition algorithms to commit financial, and personal information, or even related to political fraud. An example, in [84], shows how quickly AI helps Unilever recruiters choose suitable candidates: two first rounds are guided and judged by AI. The best are then invited to Unilever for in-person interviews, after which humans make the final hiring decisions. The final important step always requires humans, because the best candidates may be an AI! Therefore, the goal of HCAI is not to replace humans entirely but to enhance our capabilities with intelligent, human-informed technology. HCAI is beneficial but it does pose some risks. Misuse of technology, job losses, an increase in hackers and criminals, and a detrimental effect on human capabilities are the most worrisome risks. Many people think that people could become less productive, reduced creativity, and lazy as a result of advanced AI technologies. The folks believe that AI will negatively impact young people's work, home life, health, and mental health because they are addicted to smartphones and social media. HCAI faces significant challenges in the area of ethics. Because there is so much information in the world, artificial intelligence (AI) can only forecast or decide based on what it finds in the data.

Of course, HCAI plays an important role in research. If only keyword "Human centered artificial intelligence" was used in Web of Science, more than three million results have been shown in all fields. If this result was sort by categories, large number of articles focused on Computer Science AI (see Fig. 2). Very recently, [66] gives an overview of the HCAI literature. Section 5 of [1] provides a detailed analysis and summary of the relevant works in HCAI. He points out that the number of articles related to HCAI in five electronic databases, including Science Direct, Google Scholar, Springer, ACM Digital Library, and IEEE Xplore has increased steadily in recent years (period 2017-now). [5] provided a case study to illustrate how people can better classifiers, more quickly than traditional methods if human intuition and input are skillfully combined with machine learning. [2] provided a new framework using HCAI guidelines. In the research of [73], the author suggests a two-dimensional HCAI framework for enhancing AI's ability to better fulfill the needs of humans.

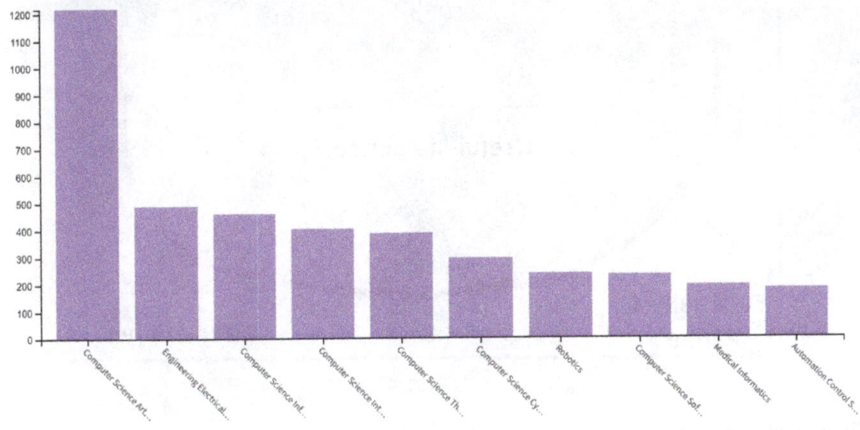

Fig. 2 Ten categories have most related work to HCAI

3 Reliability Assessment and Safety Arguments for Artificial Intelligence Systems

As mentioned in the previous sections, Artificial Intelligence (AI) systems are increasingly being entrusted with critical tasks that have significant implications on human lives and society. Therefore, it becomes necessary to ensure the reliability and safety of these systems. The main objective of this section is to provide a comprehensive overview of reliability assessment and safety arguments for AI systems. Additionally, this section will discuss various techniques and methodologies that can be employed to assess the reliability of AI systems and develop robust safety arguments.

The ability of an AI system to function in various scenarios, such as when data or user behavior changes, is referred to as reliability. According to [80], reliability can be defined mathematically as the probability that the random variable time to failure T is larger than or equal to the mission time t, or $R(t) = P(T > t)$. Why is reliability so crucial? since it has a direct bearing on human life. Car crashes claim the lives of more than a million people annually. Many lives could be saved if we could develop autonomous driving systems that are much safer and handle unexpected problems on the road with greater intelligence. Conversely, a great deal of lives will be lost if the system breaks down and people are unable to control it. Another illustration concerned the COVID-19 pandemic. The coronavirus places a significant strain on the health and medical network infrastructure. The ensuing fallout has a detrimental effect on the economy, the labor market, and the education sector. It is evident in several areas, including student health, learning capacity, and instructional style, among other things. In [86], a series of AI technologies using in the diagnosis and treatment of COVID-19 are summarized and analyzed. In addition, they proposed an AI network architecture based on edge computing and 5G which fully embodies the superior performance of AI technologies.

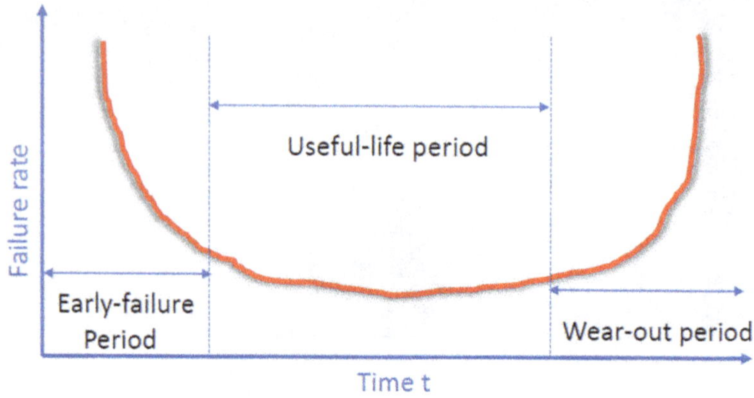

Fig. 3 Bathtub curve

A common way to measure reliability in AI systems is to use metrics such as error rate, failure rate (inverse of reliability index, MTTF-*Mean Time To Failure* or MTBF *Mean Time Between Failure* for non-repaired/repaired units corresponding), or confidence interval to measure how often the system makes mistakes or fails. These metrics assess the variability and robustness of the system. The bathtub curve is a reliability tool that is used to model the reliability of a system (see Fig. 3) Researchers use the exponential distribution, Weibull distribution law and their generalized distribution for exhibiting bathtub shaped failure rate.

Some examples of measuring the reliability of AI systems can be found in [11, 32, 35, 42, 54, 71].

A number of metrics have been proposed in the literature to evaluating the reliability and the safety of an AI system, such as the quality and quantity of the data or the algorithms and models used. In order to enhance performance and quality, it is imperative to select and implement suitable algorithms and models that align with the data and task. It has been discovered that machine learning techniques are useful for assessing the various reliability assessment model parameters. In literature, some the machine learning algorithms are very easy to interpret and efficacy. For example, three popular graph models are fault trees (FT) ([45]), decision trees (DT) ([65]), and binary decision diagrams (BDD) ([41]). There are numerous benefits and limitations ([43]). Due to different purposes and applications, one can choose the appropriate method or combine the methods to achieve better results. Many researches mention these combinations, for example see [64], the authors used BDD in FT, this provides both an accurate and efficient means of analyzing a system. Or [28] combines DT and BDD to have a new model, from then, give a decision for wind turbine manufacturing. To evaluate the effectiveness of the reliability assessment model, one can take a comparison table between the proposed model and pre-existing models. Usually, newly created models (sometimes the combination of existing—models) will overcome the previous models in reliability in making predictions or decisions ([35, 42, 54, 71]). However, there are other factors that we also need to consider, like the complexity of the model, the application environment, the cost,... Therefore, users must balance the benefits and losses to choose the appropriate/adapted model for their data.

Since reliability prediction forms the foundation of reliability engineering, it is a significant area in the reliability assessments of artificial intelligence. It is a process of estimating the probability of failure of an AI system. Reliability prediction enables developers to take proactive steps to lower risk and increase system reliability by spotting potential flaws and vulnerabilities. To predict the reliability of AI systems, developers use a variety of techniques (see [21, 36]). Fault injection is a widely used technique that involves injecting simulated faults into the system to observe how it responds to various failure scenarios. This aids in locating possible weak points and areas in need of development. Neural networks (NN) are a useful AI-based method for pattern matching ([18, 19, 68, 83]). Another method is probabilistic modeling, which calculates the probability of a system failure using statistical techniques and historical data ([36, 55, 56, 87]).

4 Safe and Reliable Machine Learning Artificial Intelligence

Safe and reliable AI systems are critical to deploying machine learning (ML) models in real-world applications. Ensuring the safety and reliability of ML systems involves addressing various challenges, such as avoiding biases, handling uncertainty, and mitigating potential risks as discussed by [6, 8, 50, 73]. Here are some key considerations for achieving safe and reliable machine learning:

- **Explainability and Interpretability**: Use interpretable models or explain complex models to enhance transparency. Understand and communicate how decisions are made by the ML model, especially in sensitive domains (see [9, 39]).
- **Robustness to Adversarial Attacks**: Design robust models to adversarial attacks, where adversaries intentionally manipulate input data to mislead the model. Employ techniques such as adversarial training to improve model resilience as discussed in [7, 16, 29, 37], etc.
- **Human-in-the-Loop (HITL)**: Incorporate human oversight in critical decision-making processes. Implement systems where human intervention can be introduced to rectify potential errors, learn from demonstrations, and refine its preferences [4, 73–75].
- **Uncertainty Quantification**: Implement methods to estimate and communicate uncertainty in predictions. Bayesian models and uncertainty calibration techniques can help quantify uncertainty [49, 77, 85].
- **Diversity and Inclusivity**: Ensure diversity in both the development team and the datasets used for training. Address potential biases related to underrepresented groups in the data [24, 69, 88].

5 Difficulties, Challenges, and Perspectives for Safety and Reliability of Artificial Intelligence Systems

The rapid development of information technology and AI's evolution has transformed industries and redefined human-machine interactions, but it is inclined to face challenges to meet ethical commitments. Security and privacy are the essential requirements of developing and deploying AI systems, which is considered the main problem. The risk of data security and privacy violation with the proliferation of AI is growing, thus requiring stronger regulations and frameworks to protect sensitive information.

5.1 Difficulties and Challenges for Deploying Safe and Reliable AI Systems

The difficulties of artificial intelligence (AI) systems encompass various aspects that pose challenges to its development, widespread adoption, and application. Some of the key difficulties and challenges include:

- **Ethical Concerns and Bias:** As AI systems become more prevalent, ethical concerns and biases have garnered significant attention. AI algorithms/systems are trained on vast datasets that can unintentionally perpetuate biases present in the data, leading to discriminatory outcomes. Therefore, the biases that arise from historical data or reflect societal biases are a major challenge in the Safety and Reliability of AI systems [57, 76].
- **Safety and Security:** Ensuring the safety and security of AI systems is a significant challenge. Vulnerabilities and potential risks, such as adversarial attacks or system failures, must be addressed to prevent unintended consequences of AI technology. Malicious actors intentionally manipulate input data to deceive or trick the AI algorithms, leading to potentially harmful consequences in sectors like cybersecurity, autonomous vehicles, and fraud detection. Ensuring the robustness and resilience of AI models against such attacks is an ongoing challenge that demands continuous innovation, rigorous testing, and the development of countermeasures [34, 40, 48].
- **Data Quality and Privacy:** AI algorithms heavily rely on large volumes of high-quality data for training and decision-making. However, acquiring and maintaining such data can be a daunting task. Issues such as data scarcity, incompleteness, and inconsistency can hinder the performance of AI systems. Additionally, concerns regarding data privacy and security arise when handling sensitive information. Striking a delicate balance between data utility and privacy is essential to build trust and comply with regulations while still extracting valuable insights from data [25, 26, 33, 61].

- **Limited Contextual Understanding:** Although AI has made remarkable strides in natural language processing and computer vision, it still struggles with contextual understanding and common-sense reasoning. AI models may misinterpret nuanced language, sarcasm, or cultural references, resulting in inaccurate results or inappropriate actions. Advancing contextual understanding requires the development of sophisticated language models and context-aware AI systems that can grasp the intricacies of human communication [15, 22, 31, 62].
- **Interpretability and Explainability:** Many AI models, particularly deep learning algorithms, operate as "black boxes." They provide outputs without clear explanations or insights into their decision-making process. This lack of interpretability and explainability can be problematic, particularly in critical applications/domains such as healthcare and finance, where transparency and accountability are key. Researchers are actively exploring methods to enhance interpretability and provide explanations for AI model outputs, aiming to build trust and facilitate human-AI collaboration (see [27, 30, 46, 82]).
- **Lack of Emotional Intelligence and Human Intuition:** Human decision-making often involves intuition, gut feelings, and instinctive judgments based on subtle cues or prior experiences. AI systems rely on data and algorithms, lacking the intuitive leaps that humans can make when faced with ambiguous or uncertain situations (see [3, 58, 79]). This limitation hinders its ability to understand and respond to complex human emotions, interpersonal dynamics, and subjective aspects of decision-making that require empathy and intuition.
- **Workforce Displacement and Economic Impact:** The widespread adoption of AI technologies can lead to workforce displacement and job market disruptions, for more details, see [17, 70, 81]. It poses challenges for reskilling and upskilling the workforce to adapt to changing job requirements and societal implications. Navigating the economic impact and ensuring a just transition are crucial considerations.

5.2 *Perspectives for Safety and Reliability of Artificial Intelligence Systems*

The above-mentioned difficulties and challenges can be addressed/ minimized. The following measures for the safety and reliability of AI systems should be considered in the future ([20, 23, 52, 59]).

- The ethical concerns and biases require transparency, accountability, and responsible AI development practices. Measures such as diverse and representative training data, algorithmic audits, ongoing monitoring, and bias mitigation techniques can help solve these issues and ensure AI technologies' fair and ethical use [23, 38].
- Rigorous testing, robust system design, continuous monitoring, and adherence to best practices help in safety and security challenges. Collaboration between AI

developers, cybersecurity experts, policymakers, and regulatory bodies is essential to establish frameworks that promote safety, security, and responsible AI deployment [14, 44].

- To tackle these challenges related to data quality and privacy, it is necessary to establish strong data governance practices, ensure the availability of diverse and inclusive datasets, embrace privacy-enhancing technologies, and foster a culture of transparency and accountability in the collection and utilization of data [59]. It is necessary to respect individuals' data privacy rights, comply with data protection regulations, and uphold ethical guidelines to ensure the responsible development and deployment of AI systems.
- Resolving these limitations requires advancements in natural language processing, machine comprehension, multimodal learning, and context-aware AI systems [10]. Ongoing research focuses on developing AI models to understand better and reason within complex contexts, bridging the gap between human-level contextual understanding and AI capabilities.
- Researchers and practitioners are developing interpretability and explainability techniques for AI systems. Methods such as model-agnostic approaches, rule extraction, attention mechanisms, and local feature importance analysis aim to provide insights into how AI models arrive at their predictions and decisions, enhancing transparency and trust in AI technology [46, 51].
- To overcome the deficiency of emotional intelligence and human intuition in AI, fostering collaboration across different disciplines, prioritizing user-centric design, and adhering to ethical development principles is essential. By integrating emotional aspects into AI systems and recognizing their wider societal implications, we can work towards creating AI technologies that possess a deeper understanding of human emotions and experiences, enabling more empathetic and responsive interactions [70].
- Society can mitigate the negative consequences of AI-driven workforce displacement and promote economic resilience by equipping individuals with the skills, resources, and support necessary to thrive in the changing job landscape [63].
- Data and AI are redefining human resources management [53]. However, research on the safety and reliability of AI systems in this field needs to be conducted in-depth to ensure privacy, fairness, as well as ethics and avoid bias in the application of AI in human resources management.

6 Concluding Remarks

Artificial Intelligence (AI) is central to the advancements of the 5IR, with a focus on humans and machines working together. Given its profound influence on modern society, the safety and reliability of AI systems are crucial. Throughout this chapter, we have explored the fundamental aspects of AI safety and reliability, focusing on the novel concept of Human-Centered Artificial Intelligence (HCAI) to emphasize the alignment of AI development with human values. We also reviewed the reliability and safety assessment methodologies, ensuring they perform consistently across different scenarios. However, recent AI systems still pose challenges, such as ethical, safety, and security, including biases, data quality issues, and the need for contextual understanding and interpretability. Overcoming these challenges requires advancements in natural language processing, privacy-preserving methods, and interdisciplinary cooperation. The safety and reliability of AI systems will receive significant attention from researchers and will rapidly evolve in the upcoming Industry 5.0. Future research will be focused on overcoming current limitations and ensuring that AI, integrated into cutting-edge technologies, can serve humanity responsibly and effectively.

References

1. Human-centered artificial intelligence, a review
2. Ahmad K, Abdelrazek M, Arora C, Baniya AA, Bano M, Grundy J (2023) Requirements engineering framework for human-centered artificial intelligence software systems. Appl Soft Comput 143:110455. ISSN 1568-4946
3. Abbasi MF, Bilal M, Rasheed K (2022) Role of human intuition in AI aided managerial decision making: a review. In: 2022 international conference on decision aid sciences and applications (DASA). IEEE, pp 713–718
4. Almada M (2019) Human intervention in automated decision-making: toward the construction of contestable systems. In: Proceedings of the seventeenth international conference on artificial intelligence and law, pp 2–11
5. Amershi S, Cakmak M, Knox WB, Kulesza T (2014) Power to the people: The role of humans in interactive machine learning. AI Mag 35(4):105–120
6. Ao S (2023) Building safe and reliable AI systems for safety critical tasks with vision-language processing. In: European conference on information retrieval. Springer, pp 423–428
7. Aquino B, Rahnama A, Seiler P, Lin L, Gupta V (2022) Robustness against adversarial attacks in neural networks using incremental dissipativity. IEEE Control Syst Lett 6:2341–2346
8. Athavale J, Baldovin A, Graefe R, Paulitsch M, Rosales R (2020) Ai and reliability trends in safety-critical autonomous systems on ground and air. In: 2020 50th annual IEEE/IFIP international conference on dependable systems and networks workshops (DSN-W). IEEE, pp 74–77
9. Balagué C (2021) The challenge of responsible AI. In: Artificial intelligence for sustainable value creation. Edward Elgar Publishing, pp 99–121
10. Basha MJ, Vijayakumar S, Jayashankari J, Alawadi AH, Durdona P (2023) Advancements in natural language processing for text understanding. In: E3S Web of Conferences, vol 399. EDP Sciences, pp 04031
11. Bhargava C, Handa M (2018) An intelligent reliability assessment technique for bipolar junction transistor using artificial intelligence techniques. Pertanika J Sci Technol 26(4):1765–1776
12. Bistouni F, Jahanshahi M (2014) Analyzing the reliability of shuffle-exchange networks using reliability block diagrams. Reliab Eng Syst Saf 132:97–106. https://doi.org/10.1016/j.ress.2014.07.012. ISSN 0951-8320
13. Cai B, Sheng C, Gao C, Liu Y, Shi M, Liu Z, Feng Q, Liu G (2023) Artificial intelligence enhanced reliability assessment methodology with small samples. IEEE Trans Neural Netw Learn Syst 34(9):6578–6590. https://doi.org/10.1109/TNNLS.2021.3128514. ISSN 2162-237X

14. Casper S, Ezell C, Siegmann C, Kolt N, Curtis TL, Bucknall B, Haupt A, Wei K, Scheurer J, Hobbhahn M et al (2024) Black-box access is insufficient for rigorous AI audits. arXiv:2401.14446
15. Anastasia C (2023) Gpt-3 and instructgpt: technological dystopianism, utopianism, and "contextualâŁž perspectives in AI ethics and industry. AI and Ethics 3(1):53–64
16. Chattopadhyay N, Chatterjee S, Chattopadhyay A (2021) Robustness against adversarial attacks using dimensionality. In: International conference on security, privacy, and applied cryptography engineering. Springer, pp 226–241
17. Chen MA, Wang JX (2024) Displacement or augmentation? the effects of AI on workforce dynamics and firm value. In: The effects of AI on workforce dynamics and firm value (April 8, 2024)
18. Cheng Y, Fu LY (2022) Nonlinear seismic inversion by physics-informed Caianiello convolutional neural networks for overpressure prediction of source rocks in the offshore Xihu depression, East China. J Petroleum Sci Eng 215 (B)
19. Chojaczyk AA, Teixeira AP, Neves LC, Cardoso JB, Guedes Soares C (2015) Review and application of artificial neural networks models in reliability analysis of steel structures. Struct Saf 52(A):78–89
20. Skalse J, Bengio Y, Russell S, Tegmark M, Seshia S, Omohundro S, Szegedy C, Goldhaber B, Ammann N et al (2024) Towards guaranteed safe AI: A framework for ensuring robust and reliable AI systems. arXiv:2405.06624
21. De P, Kar S, Ambure P, Roy K (2022) Prediction reliability of QSAR models: an overview of various validation tools. Arch Toxicol 96(5):1279–1295
22. Fernandez-Rojas R, Perry A, Singh H, Campbell B, Elsayed S, Hunjet R, Abbass HA (2019) Contextual awareness in human-advanced-vehicle systems: a survey. IEEE Access 7:33304–33328
23. Emilio F (2023) Fairness and bias in artificial intelligence: a brief survey of sources, impacts, and mitigation strategies. Sci 6(1):3
24. Fosch-Villaronga E, Poulsen A (2022) Diversity and inclusion in artificial intelligence. In: Law and artificial intelligence: regulating AI and applying AI in legal practice, pp 109–134
25. Frank E (2024) Data privacy and security in AI systems
26. Freudiger J, Rane S, Brito AE, Uzun E (2014) Privacy preserving data quality assessment for high-fidelity data sharing. In: Proceedings of the 2014 ACM workshop on information sharing & collaborative security, pp 21–29
27. Fuhrman JD, Gorre N, Hu Q, Li H, El Naqa I, Giger ML (2022) A review of explainable and interpretable AI with applications in covid-19 imaging. Med Phys 49(1):1–14
28. Garcia Marquez FP, Ramirez IS, Pliego Marugan A (2019) Decision making using logical decision tree and binary decision diagrams: a real case study of wind turbine manufacturing. ENERGIES 12(9)
29. Ghaffari Laleh N, Truhn D, Veldhuizen GP, Han T, van Treeck M, Buelow RD, Langer R, Dislich B, Boor P, Schulz V et al (2022) Adversarial attacks and adversarial robustness in computational pathology. Nat Commun 13(1):5711
30. Gilpin LH, Bau D, Yuan BZ, Bajwa A, Specter M, Kagal L (2018) Explaining explanations: an overview of interpretability of machine learning. In: 2018 IEEE 5th international conference on data science and advanced analytics (DSAA). IEEE, pp 80–89
31. Hollister DL, Gonzalez A, Hollister J (2019) Contextual reasoning in human cognition and its implications for artificial intelligence systems. Modelisation et Utilisation du Contexte, pp 1–18
32. Hong Y, Lian J, Xu L, Min J, Wang Y, Freeman LJ, Deng X (2023) Statistical perspectives on reliability of artificial intelligence systems. Quality Eng 35(1):56–78
33. Hong Y, Lian J, Xu L, Min J, Wang Y, Freeman LJ, Deng X (2023) Statistical perspectives on reliability of artificial intelligence systems. Quality Eng 35(1):56–78
34. Huang R, Zheng X, Shang Y, Xue X (2023) On challenges of AI to cognitive security and safety. Sec Saf 2:2023012

35. Huo LM, Zhu YG, Fan GF (2002) Reliability assessment of power systems by bayesian networks. In: Powercon 2002: international conference on power system technology, vols 1-4, Proceedings, pp 876–879
36. Isametova ME, Nussipali R, Martyushev NV, Malozyomov BV, Efremenkov EA, Isametov A (2022) Mathematical modeling of the reliability of polymer composite materials. Mathematics 10(21)
37. Jan STK, Messou J, Lin YC, Huang JB, Wang G (2019) Improving the robustness of adversarial attacks. Connecting the digital and physical world. Proc AAAI Conf Artif Intell 33:962–969
38. Kheya TA, Bouadjenek MR, Aryal S (2024) The pursuit of fairness in artificial intelligence models: a survey. arXiv:2403.17333
39. Kobayashi K, Alam SB (2024) Explainable, interpretable, and trustworthy AI for an intelligent digital twin: a case study on remaining useful life. Eng Appl Artif Intell 129:107620
40. Reddy Konda S (2019) Ensuring trust and security in AI: challenges and solutions for safe integration. Int J Comput Sci Technol 3(2):71–86
41. Kubica M, Opara A, Kania D (2021) Binary decision diagrams, pp 25–37. 01 2021. https://doi.org/10.1007/978-3-030-60488-2_3, ISBN 978-3-030-60487-5
42. Kuwajima H, Ishikawa F (2019) Adapting square for quality assessment of artificial intelligence systems. In: Wolter K, Schieferdecker I, Gallina B, Cukier M, Natella R, Ivaki N, Laranjeiro N (eds) 2019 IEEE 30th international symposium on software reliability engineering workshops (Issrew 2019), IEEE International symposium on software reliability engineering workshops, pp 13–18
43. Jimenez-Roa LA, Heskes T, Stoelinga M (2023) Fault trees, decision trees, and binary decision diagrams: a systematic comparison. In: Proceedings of the 31st European safety and reliability conference
44. Lapalme AN, Corbin D, Tastet O, Avram R, Hussin JG (2024) Advancing fairness in cardiac care: Strategies for mitigating bias in artificial intelligence models within cardiology. Canadian J Cardiol
45. Lee WS, Grosh DL, Tillman FA, Lie CH (1985) Fault tree analysis, methods, and applications a review. IEEE Trans Reliab R-34:194–203
46. Linardatos P, Papastefanopoulos V, Sotiris K (2020) Explainable AI: a review of machine learning interpretability methods. Entropy 23(1):18
47. Liu C, Liu X, Wu F, Xie M, Feng Y, Hu C (2018) Using artificial intelligence (Watson for oncology) for treatment recommendations amongst Chinese patients with lung cancer: feasibility study. J Med Internet Res 20(9):e11087
48. Mahanta K, Maringanti HB (2023) Safety and security in AI systems. In: Handbook of research on applications of AI, digital twin, and internet of things for sustainable development, pp 87–102. IGI Global
49. Marusich LR, Bakdash JZ, Zhou Y, Kantarcioglu M (2023) Using AI uncertainty quantification to improve human decision-making. arXiv:2309.10852
50. Patrick M, Manjul G (2021) Artificial intelligence capability: conceptualization, measurement calibration, and empirical study on its impact on organizational creativity and firm performance. Inf Manag 58(3):103434
51. Mohseni S, Zarei N, Ragan ED (2021) A multidisciplinary survey and framework for design and evaluation of explainable AI systems. ACM Trans Interact Intell Syst (TiiS) 11(3–4):1–45
52. Myllyaho L, Raatikainen M, Mannistö T, Mikkonen T, Nurminen JK (2021) Systematic literature review of validation methods for AI systems. J Syst Softw 181:111050
53. Nawaz N, Arunachalam H, Pathi BK, Gajenderan V (2024) The adoption of artificial intelligence in human resources management practices. Int J Inf Manag Data Insights 4(1):100208
54. Nguyen HD, Tran KP, Zeng X, Koehl L, Tartare G (2020) An improved ensemble machine learning algorithm for wearable sensor data based human activity recognition. In: Reliability and statistical computing. Springer, pp 207–228
55. Niu XP, Wang RZ, Liao D, Zhu SP, Zhang XC, Keshtegar B (2021) Probabilistic modeling of uncertainties in fatigue reliability analysis of turbine bladed disks. Int J Fatigue 142

56. Jakub N, Rafal W (2018) Recent advances in electricity price forecasting: a review of probabilistic forecasting. Renew Sustain Energy Rev 81(1):1548–1568
57. Ntoutsi E, Fafalios P, Gadiraju U, Iosifidis V, Nejdl W, Vidal ME, Ruggieri S, Turini F, Papadopoulos S, Krasanakis E et al (2020) Bias in data-driven artificial intelligence systems-an introductory survey. Wiley Interdiscip Rev: Data Min Knowl Dis 10(3):e1356
58. Omafume O (2023) Human intelligence versus AI: implications for emotional aspects of human communication. J Adv Res Soc Sci 6(2):76–85
59. Park S, Gil-Garcia JR (2022) Open data innovation: visualizations and process redesign as a way to bridge the transparency-accountability gap. Gov Inf Quart 39(1):101456
60. Patecornell M (1984) Fault-trees vs event trees in reliability-analysis. Risk Anal 4(3):177–186. https://doi.org/10.1111/j.1539-6924.1984.tb00137.x, ISSN 0272-4332
61. Petersson L, Larsson I, Nygren JM, Nilsen P, Neher M, Reed JE, Tyskbo D, Svedberg P (2022) Challenges to implementing artificial intelligence in healthcare: a qualitative interview study with healthcare leaders in Sweden. BMC Health Serv Res 22(1):850
62. Pichler M, Bodenhofer U, Schwinger W (2004) Context-awareness and artificial intelligence. ÖGAI J 23(1):4–11
63. Popoola OA, Akinsanya MO, Nzeako G, Chukwurah EG, Okeke CD (2024) The impact of automation on maritime workforce management: a conceptual framework. Int J Manag Entrepreneurship Res 6(5):1467–1488
64. Reay KA, Andrews JD (2002) A fault tree analysis strategy using binary decision diagrams. Reliab Eng Syst Saf 78(1):45–56
65. Rokach L (2008) Data mining with decision trees: theory and applications. Series in machine perception and artificial intelligence. ISBN: 9789812771728
66. Schmager S, Pappas I, Vassilakopoulou P (2023) Defining human-centered AI: a comprehensive review of HCAI literature
67. Schmidt A (2020) Interactive human centered artificial intelligence: a definition and research challenges. In: Proceedings of the international conference on advanced visual interfaces, pp 1–4
68. Shafiq A, Çolak AB, Lone SA, Sindhu TN, Muhammad T (2024) Reliability modeling and analysis of mixture of exponential distributions using artificial neural network. Math Methods Appl Sci 47(5, SI):3308–3328
69. Shams RA, Zowghi D, Bano M (2023) AI and the quest for diversity and inclusion: a systematic literature review. AI and Ethics, pp 1–28
70. Shen Y, Zhang X (2024) The impact of artificial intelligence on employment: the role of virtual agglomeration. Human Soc Sci Commun 11(1):1–14
71. Sheptunov SA, Larionov MV, Suhanova NV, Salakhov MR, Solomentsev YM, Il'ya SK (2016) Simulating reliability of the robotic system software on the basis of artificial intelligence. In: Proceedings of the 2016 IEEE conference on quality management, transport and information security, information technologies (IT & MQ & IS). IEEE. ISBN 978-1-5090-3680-6. IEEE conference on quality management, transport and information security, information technologies (IT & MQ & IS), Nalchik, RUSSIA, OCT 04-11, 2016
72. Shneiderman B (2020) Bridging the gap between ethics and practice: guidelines for reliable, safe, and trustworthy human-centered AI systems. ACM Trans Interact Intell Syst (TIIS) 10(4):1–31
73. Shneiderman B (2020) Human-centered artificial intelligence: three fresh ideas. AIS Trans Hum-Comput Interact 12(3):109–124
74. Shneiderman B (2022) Human-centered AI. Oxford University Press
75. Sterz S, Baum K, Biewer S, Hermanns H, Lauber-Ronsberg A, Meinel P, Langer M (2024) On the quest for effectiveness in human oversight: Interdisciplinary perspectives. arXiv:2404.04059
76. Niral S (2022) Bias and ethical concerns in machine learning. ISACA J 4(1):1–4
77. Toth Z, Schultz P, Mullen S, Demargne J, Zhu Y (2007) Completing the forecast: assessing and communicating forecast uncertainty. In: ECMWF workshop on ensemble prediction, pp 7–9

78. Towhidi F, Lashkari AH, Hosseini RS. (2009) Binary decision diagram (bdd). In: International conference on future computer and communications, proceedings, pp 496+
79. Turan N, Aydın GO, Özsaban A, Kaya H, Aksel G, Yılmaz A, Hasmaden E, Akkuş Y (2019) Intuition and emotional intelligence: a study in nursing students. Cogent Psychol 6(1):1633077
80. Verma AK, Ajit S, Karanki DR (2016) Reliability and safety engineering: Second Edition. 01 2016. ISBN 978-1-4471- 6268-1
81. Paolo V, Kirsten S (2021) The impact of forced displacement on host communities: a review of the empirical literature in economics. J Dev Econ 150:102606
82. Bhuman V (2023) Explainable AI: Assessing methods to make AI systems more transparent and interpretable. Int J New Media Stud: Int Peer Rev Schol Indexed J 10(1):236–242
83. Wang S, Ren P, Takyi-Aninakwa P, Jin S, Fernandez C (2022) A critical review of improved deep convolutional neural network for multi-timescale state prediction of lithium-ion batteries. Energies 15(14)
84. Wilson HJ, Daugherty PR (2018) Collaborative intelligence: humans and AI are joining forces. Harvard Bus Rev 96(4):114–123
85. Wörtwein T, Morency, L-P (2020) Simple and effective approaches for uncertainty prediction in facial action unit intensity regression. In: 2020 15th IEEE international conference on automatic face and gesture recognition (FG 2020). IEEE, pp 452–456
86. Zhang P, Wang C, Kumar N, Jiang C, Lu Q, Choo KK, Rodrigues JJ (2021) Artificial intelligence technologies for covid-19-like epidemics: methods and challenges. IEEE Netw 35(3):27–33
87. Zhu SP, Huang HZ, Peng W, Wang HK, Mahadevan S (2016) Probabilistic physics of failure-based framework for fatigue life prediction of aircraft gas turbine discs under uncertainty. Reliab Eng Syst Saf 146:1–12
88. Zowghi D, da Rimini F (2023) Diversity and inclusion in artificial intelligence. arXiv:2305.12728

www.ingramcontent.com/pod-product-compliance
Lightning Source LLC
Chambersburg PA
CBHW071645250825
31638CB00004B/185
9783031714948